Decarbonize Public and Commercial Buildings

Yi Jiang · Shan Hu · Xiaohua Liu · Tao Zhang · Qingpeng Wei

Decarbonize Public and Commercial Buildings

China Building Energy and Emission Yearbook 2022

Building Energy Research Center Tsinghua University

Springer

Yi Jiang
Building Energy Research Center
Tsinghua University
Beijing, China

Shan Hu
Building Energy Research Center
Tsinghua University
Beijing, China

Xiaohua Liu
Building Energy Research Center
Tsinghua University
Beijing, China

Tao Zhang
Building Energy Research Center
Tsinghua University
Beijing, China

Qingpeng Wei
Building Energy Research Center
Tsinghua University
Beijing, China

https://doi.org/10.1007/978-981-19-5526-6

This Springer imprint is published by the registered company Springer Nature Singapore Pte Ltd.
The registered company address is: 152 Beach Road, #21-01/04 Gateway East, Singapore 189721, Singapore
(www.springernature.com/mycopy)

Preface

As United Nations Secretary-General António Guterres said: The impact of the coronavirus is both immediate and dreadful. But there is another deep emergency—the planet's unfolding environment. Fulfilling the low-carbon strategy is not only to address the challenge of climate change but also a new mode of the human development under the concept of ecological civilization. Only in this way can we wean the world off its dependence on fossil fuels and achieve sustainable development goals.

In order to achieve the peak carbon dioxide emission and carbon neutrality strategy, many projects are actively promoted nationwide. The ensuing energy revolution will bring great changes to our development, economy, and people's life. It is a great opportunity and challenge for the Chinese people to achieve the goals of this energy revolution while realizing China's national rejuvenation.

Public and commercial building sector is an important battlefield to realize building low-carbon goals. Low-carbon development strategy has proposed many new subjects and new tasks for the development of P&C buildings, which are worthy of research. From the perspective of building energy conservation, we proposed reducing the energy demand by passive design of architecture and improving the efficiency of energy use by active optimization of electromechanical system. Nowadays, a requirement should be added on the basis of the passive design of architecture: how to optimize the building construction to make full use of its own surface resources in the utilization of building photovoltaic, so as to transform the buildings from energy consumers to energy prosumers? For electromechanical systems, there are four requirements: complete electrification to eliminate fossil fuels; improve system efficiency to reduce energy consumption; promote decentralized methods to avoid excess supply; develop flexible power to achieve effective consumption of wind power and solar power. These 2+4 requirements mentioned above are the basic principles for P&C buildings to achieve low-carbon goals. At present, some new technologies based on these basic principles are under research and pilot demonstration.

This publication summarizes and analyzes the current situation of the stock, energy consumption, and carbon emissions of China's public and commercial buildings. It is expected to provide data and references for building and system engineers and policymakers on the topic of the energy conservation and low-carbon development of China's P&C buildings. At the same time, it is also available for international counterparts to understand the development of P&C buildings in China and jointly promote energy conservation and low-carbon development of the global building sector.

2022 is coming. I hope that, this year, we can all stay away from COVID-19, and the world and China will become better.

Beijing, China
March 2022

Yi Jiang

Acknowledgement

This publication was prepared by the Building Energy Research Center (BERC) of Tsinghua University.

The lead authors were Dr. Shan Hu, Professor Yi Jiang, Professor Xiaohua Liu, Dr. Tao Zhang, and Professor Qingpeng Wei. Other authors also contributed important sections to this report. The authors' list of report is given below:

Dr. Shan Hu (Chaps. 1, 2, Sect. 3.1)
Prof. Yi Jiang (Chaps. 2, 3)
Professor Xiaohua Liu (Chaps. 3, 4)
Dr. Tao Zhang (Chaps. 3, 4)
Professor Qingpeng Wei (Sects. 3.1 and 3.2)
Professor Baolong Wang (Sect. 2.3),
Yang Zhang (Sects. 2.2, 2.3 and 3.1)
Ziyi Yang (Sects. 2.4 and 3.1)

The research of this report was supported by the Youth Program of the National Natural Science Foundation of China (Grant No. 51908311), the National Key Research Program of China (No. 2022YFC3802500), and National Natural Science Foundation of China (No. 52278114).

Building Energy Research Center of Tsinghua University

The Building Energy Research Center (BERC) of Tsinghua University was founded in 2005 by Professor Yi Jiang and his colleagues. The mission of BERC is to be devoted to the development of energy-efficient and environmentally responsible buildings in China in accordance with national and international energy and environmental targets, including buildings research and innovation.

The principal research activities within BERC include:

- Assessment of the current buildings status in China and the provision of strategic outlooks on buildings energy consumption and efficiency.
- Occupant behavior and building simulation research.
- Research and development (R&D) of innovative high-efficiency buildings technology and systems.
- Energy efficiency application research on subsectors, including: space heating in Northern China; rural residential buildings and urban residential buildings; and public and commercial buildings.

Since 2007, BERC has published *Annual Report on China Building Energy Efficiency (in Chinese)*, to provide data references, technical and policy suggestions to policy makers and engineers in the building energy conservation sector. BERC is also involved in international exchange and cooperation, including on-going collaboration with the International Energy Agency.

Executive Summary

Building Construction Speed in China Has Slowed Down

In 2020, China's urbanization rate rise from 37.7% in 2001 to 63.9% in 2020. Rapid urbanization drives the continuous development of the construction sector. From 2014 to 2019, the newly built building stock of civil buildings in China slowly decreased year by year. In 2020, the newly built building stock of civil buildings dropped to 3.8 billion m^2 due to the impact of COVID-19. In 2020, the total building stock in China was about 66 billion m^2, including urban residential buildings accounted for 29.2 billion m^2, rural residential buildings accounted for 22.7 billion m^2, and public and commercial (P&C) buildings accounted for 14 billion m^2. The floor area for northern urban heating was 15.6 billion m^2.

Compared with developed countries, the per capita floor area of residential buildings in China is close to that in developed countries such as the EU and Japan. But the per capita floor area of public and commercial buildings in China is still relatively low. Hospitals, schools, transportation hubs, and service buildings will be the main items for new building construction in the future.

Transformation of Building Construction Mode is Ongoing

Building construction-related embodied energy in China was 0.52 Gtce in 2020, accounting for 10% of China's total energy consumption. The embodied energy uses of urban residential, rural residential, and P&C buildings accounted for 71%, 6%, and 23%, respectively.

In 2020, the total carbon emissions from civil building construction in China was about 1.5 $GtCO_2$, mainly including carbon emissions from manufacturing and transportation of building materials (77%), emission from cement industrial process (20%), and building on-site construction (3%).

The fast pace of urbanization in China drove the demand for building materials. As urbanization and infrastructure development have achieved initial progress in China, the transformation of the construction mode is ongoing. For the next round of urbanization, the demolition of buildings that have not reached the end of life shall be abolished. Technologies to extend the building life cycle shall be invented. Buildings and infrastructures should be properly repaired, and the building life cycle should be extended to facilitate industry transformation and total energy and emission control.

Building Operation Energy and Emission Increasing Steadily

The total commercial energy consumption of building operations was 1.06 Gtce, accounting for 21% of total primary energy consumption in China. Energy use intensity of public and commercial buildings was highest among the four energy use categories and continued to increase.

In 2020, the total carbon emissions during building operations in China was 2.18 billion tCO_2, the carbon emission intensity was 1.5 t/cap and 33 kg/m^2.

Total building energy use and emissions in China are high, but their intensity is still very low. A global comparison of building energy consumption and emissions shows that China needs to find a unique pathway to achieve sustainable and decarbonized development in the building sector.

The shares of direct carbon emissions, indirect carbon emissions from electricity use, and indirect carbon emissions from heating in total building carbon emissions were 27%, 52%, and 21%, respectively. Direct carbon emissions have already peaked, while indirect emissions from electricity and heating may peak by 2030. Technological innovations and appropriate policy instruments are needed to reach the peak of carbon emissions and carbon neutrality in the building sector. Electrification of energy use in buildings is key to reducing direct emissions in buildings. PEDF (photovoltaic, energy storage, direct current, and flexibility) buildings and a zero-carbon heating system in northern China are critical to reducing indirect emissions from buildings.

Carbon Neutrality Goal Gives Public Building New Role than Energy Saving

Energy use intensity and carbon emission intensity of public and commercial buildings are the highest among civil buildings. To decarbonize public and commercial buildings, passive measures must be taken first to reduce demand and then active measures to optimize efficiency. Passive technologies used in the design and construction of buildings minimize the demands of buildings for cooling, heating, and artificial daylighting. The electromechanical system of each building should follow the principle of "electrification, distributed, high efficiency, and flexible electricity consumption."

The goal of carbon neutrality drives the organic unification of low-carbon and energy-saving public building goals. Energy conservation is an important foundation for achieving the low-carbon goal, while the low-carbon or even zero-carbon goal sets new and higher requirements for building energy conservation in buildings. The new power distribution systems of the PEDF buildings are expected to enable flexible energy use of buildings to better support the whole energy system or even the whole society in achieving the low-carbon target.

Contents

1 Introduction 1

2 China's Building Energy Use and GHG Emissions 3
2.1 Basic Situation of China's Building Sector 3
2.1.1 Urban and Rural Demographic 3
2.1.2 Building Stock 4
2.2 Energy Consumption of China's Building Sector 6
2.2.1 Building Operation Energy Consumption 8
2.2.2 Embodied Energy of Building Sector 17
2.3 GHG Emissions of China's Building Sector 21
2.3.1 Carbon Dioxide Emissions from Building Operation Energy Consumption 22
2.3.2 Embodied CO_2 Emission of Building Sector 26
2.3.3 Non-CO_2 GHG Emissions in the Building Sector 29
2.4 Energy Consumption and GHG Emissions in the Global Building Sector 33
2.4.1 Energy Consumption by Global Building Operation 33
2.4.2 Boundaries and Comparative Study Methods of Building Energy Consumption and Emissions 35
2.4.3 Global Energy Consumption and Carbon Emissions in the Building Sector 38

3 Energy and Emissions of Public and Commercial Buildings 45
3.1 Trends in the Development of Public and Commercial Building Stock and Type 45
3.2 Characteristics and Development Trend of the Energy Consumption of Public and Commercial Buildings 49
3.2.1 Changes in Key Fields of Public and Commercial Building Energy Saving During the 13th Five-Year Plan Period 49
3.2.2 Characteristics of Changes in Energy Use Intensity of Main Types of P&C Buildings 52

3.2.3 Changes in the "Bimodal Distribution" Characteristics of Energy Use Intensity of P&C Buildings 59

4 Decarbonize Public and Commercial Buildings 63
4.1 Role of Public Buildings in Achievement of the "Dual Carbon" Goals 63
4.2 Are Public Buildings Still Need to Be Energy-Saving in Addition to Low-Carbon? 72
4.3 CO_2 Concentration and Ventilation for Epidemic Prevention 78
4.4 Further Understanding on the Contribution of Public Building Energy Systems to the "Dual Carbon" Goals 90
4.5 Further Understanding of District Energy Systems Oriented Toward the "Dual Carbon" Goals 97

List of Figures

Fig. 2.1 Population growth in China by year (2001–2020) 4
Fig. 2.2 Newly built and demolished building stock of civil buildings in China (2007–2020) 5
Fig. 2.3 Newly built building stock of P&C buildings according to different functions (2001, 2020) 5
Fig. 2.4 China's existing building stock (2001–2020). *Source* Estimation results from CBEEM of the Building Energy Research Center (BERC), Tsinghua University. The newly built building stock entered in the model is the data under the statistical standards for construction enterprises as specified in the China Statistical Yearbook on Construction 6
Fig. 2.5 Comparison of per capita building floor area between China and other countries 7
Fig. 2.6 Boundaries and types of energy consumption and GHG emissions in the building sector 7
Fig. 2.7 Primary energy consumption and total electricity consumption of building operation in China (2010–2020) 11
Fig. 2.8 Building operation energy consumption in China (2020) 12
Fig. 2.9 Changes in total building energy use and energy use intensity by year (2010–2020) 12
Fig. 2.10 Embodied energy use of China's civil buildings (2004–2020). *Source* Estimation by BERC, Tsinghua University. Civil building construction is included only[1] 17
Fig. 2.11 Embodied energy consumption of China's construction sector (2004–2020).[2] *Source* Estimation by BERC, Tsinghua University. The construction sector involves the construction of civil buildings, production buildings and infrastructures 18
Fig. 2.12 Comparison of total energy use and energy consumption per unit of value added in manufacturing[1] 18

Fig. 2.13 Comparison of industry structure for energy use in the manufacturing sector between China and some developed countries[1] ... 19
Fig. 2.14 Comparison of energy consumption per unit of value added among manufacturing sub-sectors in China (2017)[1] ... 19
Fig. 2.15 Energy consumption for building material production[1] ... 20
Fig. 2.16 Carbon dioxide emissions from building operation (2020) ... 22
Fig. 2.17 Carbon dioxide emissions from building operation in China (2020) ... 25
Fig. 2.18 Carbon emissions of building energy use (2010–2020) ... 26
Fig. 2.19 Embodied carbon emissions of civil buildings (2004–2020).[1] *Source* Estimation by BERC, Tsinghua University. Civil building construction is included only ... 27
Fig. 2.20 Embodied carbon emissions of the construction sector in China (2004–2020). *Source* Estimation by BERC, Tsinghua University. The construction sector involves the construction of civil buildings, production buildings and infrastructures ... 27
Fig. 2.21 Terminal energy use and CO_2 emissions in the global building sector (2020). *Source* 2021 Global Status Report for Buildings and Construction, International Energy Agency. The construction sector involves civil building construction, production building construction, and infrastructure construction. This figure uses the terminal energy consumption data provided by IEA, which is obtained through direct addition of heat consumption for heating, electricity consumption of buildings and terminal use of various energy types. The electricity consumption is converted to primary energy using a calorific value equivalent method. This conversion method is different from that used in the comparison of building energy consumption among countries in the ensuing part of this research, so it should be treated differently in data comparison ... 34
Fig. 2.22 China's building energy use and CO_2 emissions (2020). *Source* Estimation with CBEEM of BERC, Tsinghua University. The construction sector involves the construction of civil buildings, production buildings and infrastructures ... 35

Fig. 2.23 Comparison of building operation energy consumption in different countries. *Source* CBEEM of BERC, Tsinghua University; World Energy Balances, Energy Efficiency Indicators database (2021 edition) of IEA; WDI database of the World Bank; Satish Kumar (2019) of India.[1] Data of 2020 is adopted for China, data of 2018 for Canada and Sweden, and data of 2019 for other countries. In the total energy consumption of buildings, the electricity consumption of different countries is converted to primary energy consumption based on China's coefficient of coal consumption for thermal power supply 38
Fig. 2.24 Comparison of building operation electricity and non-electricity use in different countries. *Source* Same as above 39
Fig. 2.25 Electrification rate of energy use in building sector in different countries (2001–2019) 40
Fig. 2.26 Comparison of per capita carbon emissions of different countries (2019). *Source* Data of countries in 2019 as provided in the CO_2 Emissions from the Fuel Combustion Highlights 2020 database, IEA. Data of China are the results of 2020 as estimated with CBEEM of BERC, Tsinghua University 41
Fig. 2.27 Comparison of carbon emissions from building operation in different countries (2019). *Source* Data of countries in 2019 as provided in the CO_2 Emissions from the Fuel Combustion Highlights 2020 database, IEA. Data of China are the results of 2020 as estimated with CBEEM of BERC, Tsinghua University 42
Fig. 2.28 Trajectory of building operation carbon emission in different countries (2000, 2019). *Source* Data of countries from 2000 to 2019 as provided in the CO_2 Emissions from the Fuel Combustion Highlights 2020 database, IEA. Data of China are the results of 2020 as estimated by CBEEM of BERC, Tsinghua University 42
Fig. 3.1 Growth trend of building stock of public and commercial buildings by type 46
Fig. 3.2 Trend of change in newly built building stock of public and commercial buildings by type 46
Fig. 3.3 Numbers and annual increases of urban rail transit stations in operation 47
Fig. 3.4 Comparison of per capita P&C building floor area in different countries 47

Fig. 3.5 Comparison of per capita floor area of different types of P&C buildings in different countries. *Notes* [1] For the office building floor area data of Japan, the original data indicators are related to "office + shop", so this type includes office and mall buildings. As a result, the statistical standard of Japan is different from that of other countries. [2] The P&C buildings in the US fall into 15 categories, and the data of some types of P&C buildings are merged according to the classification method used in the figure 48
Fig. 3.6 Monthly electricity consumption of connected commercial buildings in 2019–2020 54
Fig. 3.7 Monthly electricity consumption of different types of connected commercial buildings in 2019–2020 55
Fig. 3.8 Changes in electricity consumption of connected public buildings over the years during the 13th Five-Year Plan period 55
Fig. 3.9 Changes in electricity consumption of different types of connected public buildings over the years during the 13th Five-Year Plan period 56
Fig. 3.10 Electricity consumption of monitored building in Shenzhen 57
Fig. 3.11 Comparison of electricity consumption of monitored public building in Shenzhen from 2019 to 2020 57
Fig. 3.12 Electricity consumption intensity of monitored public buildings 58
Fig. 3.13 Energy use intensity distribution 60
Fig. 4.1 Changes over time in generated outputs of wind and solar PV power (with a typical week of July in Beijing as an example) 64
Fig. 4.2 Equilibrium diagram of future electric power systems 67
Fig. 4.3 Examples of public buildings using PV 70
Fig. 4.4 Interaction between public buildings and future low-carbon electric power system 71
Fig. 4.5 Mismatch between PV renewable energy and building energy consumption 77
Fig. 4.6 Relationship between indoor CO_2 concentration and per capita fresh air volume 80
Fig. 4.7 Main indicators considered in indoor environment monitoring and operation 82
Fig. 4.8 Relationship of heat, humidity and pollution sources between indoors and outdoors 83
Fig. 4.9 Relation between airborne transmission risk and ventilation rate[3]: **a** $q = 14$; **b** $q = 48$ 88
Fig. 4.10 Public building energy demands and main supply modes 91
Fig. 4.11 Hourly normalized electricity consumption curves of public buildings in Shanghai (a typical day in summer)[1] 93

Fig. 4.12 Cooling, heating and electricity demands of district buildings and main forms of external inputs ... 104
Fig. 4.13 Future district buildings to be involved in power grid response as flexible load ... 106

List of Tables

Table 2.1 Building operation energy consumption in China in 2020 11
Table 2.2 HCFCs and HFCs refrigerants commonly used at present in China 29
Table 2.3 GWP values of common refrigerants 30
Table 4.1 Installed capacity and generating capacity of China's electric power systems in 2019 and 2050 64
Table 4.2 Peak-to-valley ratios of daily electricity consumption of public buildings in Shanghai (a typical day in summer)[2] 94

Chapter 1
Introduction

Founded in 2005, the Building Energy Research Center of Tsinghua University (BERC), has worked continuously on China's buildings sector, to comprehensively analyses the current status, unique features and key issues of China's buildings energy use. The aim of BERC's research is to understand and reveal the experiences and lessons from global buildings energy comparison, and then to recommend future energy technology and perspectives on China's buildings energy conservation work as well as provide best practice project demonstration.

Since 2007, BERC has published *Annual Report on China Building Energy Efficiency (in Chinese)*, flagging this on-going research work and key data for China's buildings sector. Those annual reports (2007–2022) show key data for China's buildings energy use, discuss key issues of technical developing trends, indicate potential policy and technology perspectives and demonstrate best practices in the buildings sector. As China has become increasingly important to global energy strategy and climate change, greater attention both nationally and internationally has been paid to China's buildings sector.

From 2016, BERC released an English version of this annual report, for international researchers who are interested to know about China's building sector.

In this report, China's building energy and emission status and trend were discussed and updated every year. Four categories of China's building energy use were defined according to their own characteristics, and every year one category was selected to the topic. In 2022, the topic was public and commercial buildings. Therefore, this report includes three parts.

Chapter 2 updated China's building energy use and emissions during construction phase and operation phase in 2020. Detailed data on China's building sector including population, building stock, total amount and intensity of energy and emissions could be found in this chapter.

Y. Jiang et al., *Decarbonize Public and Commercial Buildings*,
https://doi.org/10.1007/978-981-19-5526-6_1

Chapter 3 is the section focus on one specific building energy use category, which is public and commercial buildings. This chapter described developing trend of building stock, type, energy, and emissions of public and commercial buildings.

Chapter 4 discussed concept and requirement for public and commercial buildings towards carbon neutrality target, and decarbonization pathways for public and commercial buildings, and district energy system.

Chapter 2
China's Building Energy Use and GHG Emissions

2.1 Basic Situation of China's Building Sector

2.1.1 Urban and Rural Demographic

In recent years, urbanization has grown rapidly in China. In 2020, China had an urban population of 902 million and a rural population of 510 million, with the urbanization rate rising from 37.7% in 2001 to 63.9%, as shown in Fig. 2.1.

Urbanization is fundamentally characterized by the massive migration of people from rural areas to cities. In China, urbanization means that the people, for the most part, migrate to superlarge cities and county-level cities. According to relevant research by Li Xiaojiang, former director of the China Academy of Urban Planning and Design, from 2000 to 2010, 41% of urban population growth was contributed by megacities, superlarge cities and large cities, and 37% came from counties and towns. Recent years have seen a significant decline in the population growth of overpopulated large cities with excessively strict entry restrictions. For instance, the number of permanent residents in Beijing and Shanghai have remained basically stable since 2016.

On the other hand, rural residents migrate to counties and small towns, which is another characteristic of urbanization in China. Currently, about one-fourth of Chinese people now lives in small towns. Until 2020, China had a total of 1495 counties, with a total population of 140 million in built-up areas, and 18,822 designated towns, with a total population of 184 million in built-up areas. Since 2001, the residential building stock in designated towns has doubled, from 2.86 to 6.14 billion m^2. Historically, the main function of such small towns with a population of 10–100 thousand has been to provide surrounding rural areas with economic and trade, cultural, and medical services, and the support of their economic operation has been determined by the scale of surrounding agriculture, animal husbandry and forestry they serve. The economic activities of these small towns have been dominated by the service sector, and it has been very difficult to arrange activities of the

Y. Jiang et al., *Decarbonize Public and Commercial Buildings*,
https://doi.org/10.1007/978-981-19-5526-6_2

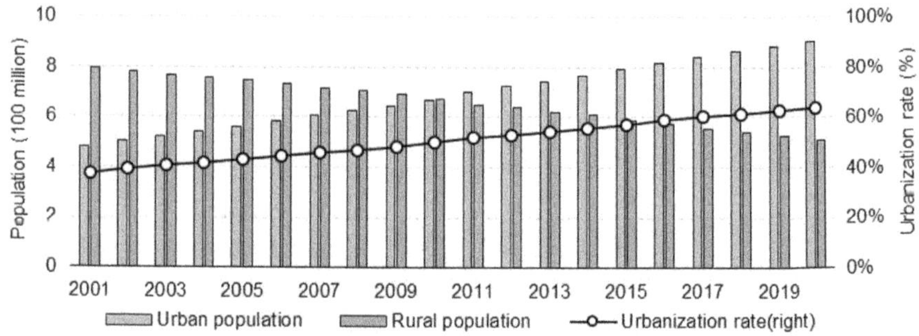

Fig. 2.1 Population growth in China by year (2001–2020)

secondary industry there. As the proportion of urbanization rises and the agricultural population declines, the corresponding service functions of small towns also decrease accordingly. Housing vacancy occurs in these small towns when the ratio of their population to the surrounding rural population they serve is higher than a certain limit. How to plan the functions of these small towns and develop their infrastructure system represented by energy system in a scientific and rational way to enable their sustainable development will be an important issue to be addressed in the new era. These small towns have seldom developed into manufacturing bases, but could become the places of residence for a multitude of aged people, women and children currently left behind in rural areas to meet their demands for low-cost and high-quality places of residence, change the phenomenon that the rural inhabitants are dominated by a large number of "993861" non-producers (i.e. the elderly, women and children), and provide such left-behind population with better social security and education services. Considering these functions and development goals, the housing construction patterns, infrastructure construction modes, and especially energy system building methods in small towns should be different from those in large cities and in scattered natural villages. This should be given enough attention.

2.1.2 *Building Stock*

Rapid urbanization drives continuous development of the construction sector, and the scale of China's construction sector has been expanding. From 2007 to 2020, thanks to the rapid growth of building construction in China, the floor space greatly expanded in urban and rural areas. Specifically, from 2007 to 2014, the newly built building stock for civil buildings grew steadily from 2 billion m^2 per annum to more than 4 billion m^2. From 2014 to 2019, the newly built building stock of civil buildings in China slowly decreased year by year, but basically remained above 4 billion m^2. In 2020, due to the impact of COVID-19, the newly built building stock of civil buildings dropped to 3.8 billion m^2. The total newly built building stock of urban residential buildings and public and commercial (P&C) buildings fell from about 3.6

billion m^2 in 2014 to 3.34 billion m^2 in 2020 (Fig. 2.2). Driven by the large number of construction projects, the demolished building stock of urban residential areas and P&C buildings increased rapidly from 700 million m^2 in 2007 to a stable level of approximately 1.6 billion m^2 per year at present.

In 2020, residential buildings and non-residential buildings made up about 78 and 22% of the newly built building stock of civil buildings in China respectively. According to the difference in building functions, public and commercial (P&C) buildings can be categorized into offices, hotels, malls, hospitals, schools and others. From 2001 to 2020, the main types completed each year were dominated by offices, malls and schools. In 2020, the total newly built building stock of these three types accounted for about 72% of that of public and commercial (P&C) buildings, with the proportions of malls, office buildings and schools at 31%, 22% and 19% respectively. For the remaining types, hospitals and hotels had a smaller proportion of 6% and 3% respectively (Fig. 2.3).

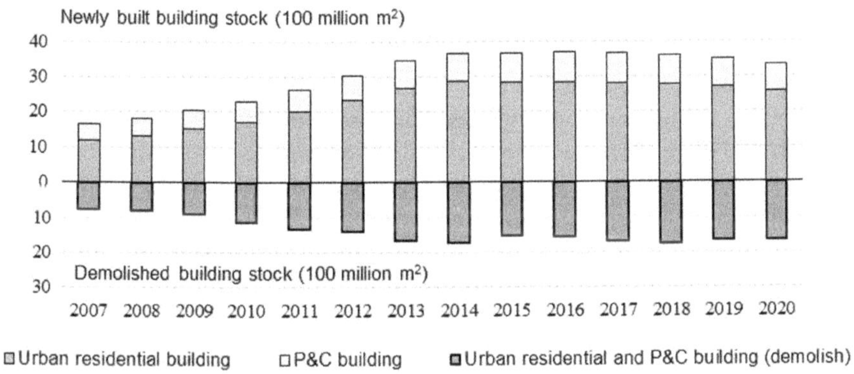

Fig. 2.2 Newly built and demolished building stock of civil buildings in China (2007–2020)

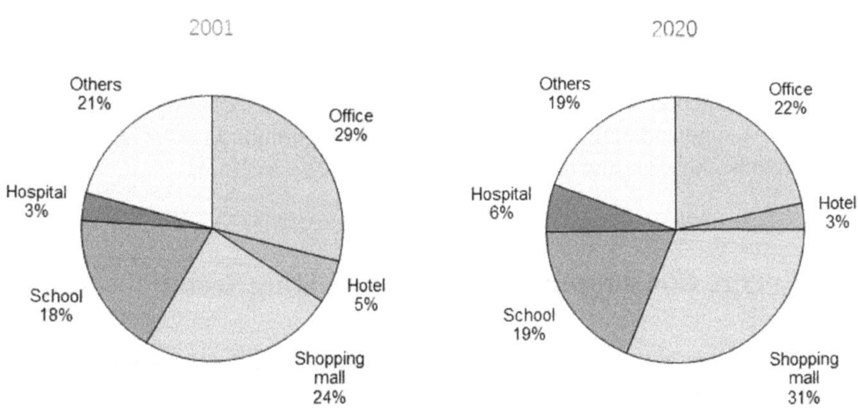

Fig. 2.3 Newly built building stock of P&C buildings according to different functions (2001, 2020)

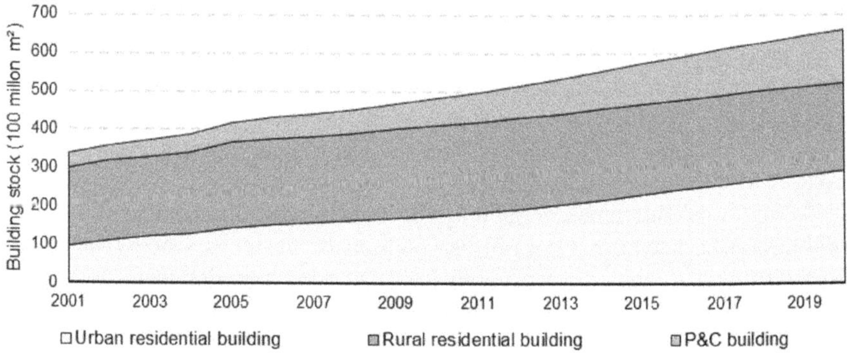

Fig. 2.4 China's existing building stock (2001–2020). *Source* Estimation results from CBEEM of the Building Energy Research Center (BERC), Tsinghua University. The newly built building stock entered in the model is the data under the statistical standards for construction enterprises as specified in the China Statistical Yearbook on Construction

Large-scale building construction activity has led to the rapid growth of China's building stock every year. In 2020, the total building stock in China was about 66 billion m^2, including urban residential buildings accounted for 29.2 billion m^2, rural residential buildings accounted for 22.7 billion m^2, and P&C buildings accounted for 14 billion m^2 (Fig. 2.4). The floor area for northern urban heating stood at 15.6 billion m^2.

According to the comparison of per capita building floor area between China and other countries, as shown in Fig. 2.5, the per capita residential building floor area in China is close to that in some developed countries, but the per capita P&C building floor area in China is still lower than that in some developed countries. For the existing P&C buildings in China, the per capita office building floor area is relatively reasonable, but the per capita areas of malls, hospitals and schools are still at a lower level. With the rapid development of e-commerce, it is very difficult for the scale of malls to continue to grow. However, there is still potential for growth for the scale of such public service buildings as hospitals and schools. Therefore, these public service buildings could be the main items for the next round of new P&C building construction in China. In addition, other buildings including transportation hubs, recreational and athletic buildings as well as community center buildings are expected to be the mainstay of future P&C building construction.

2.2 Energy Consumption of China's Building Sector

Energy use and emissions for the building sector covered different phases of buildings, including building construction, operation, and demolition. However, in China, the majority of energy use and GHG emissions for the building sector take place in building construction and operation, which are the focus of this report, as shown in Fig. 2.6.

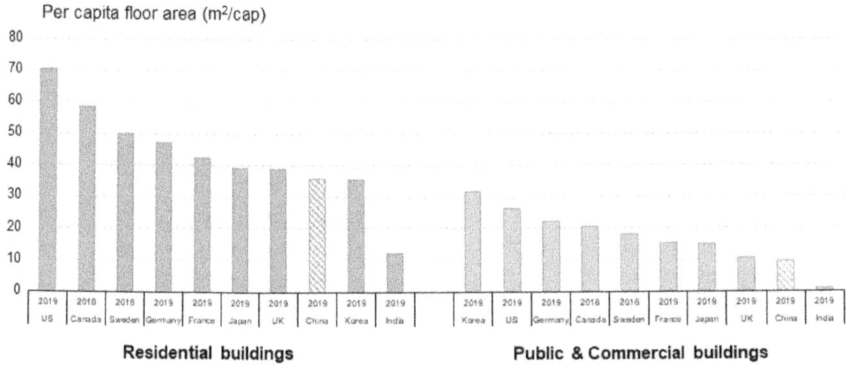

Fig. 2.5 Comparison of per capita building floor area between China and other countries

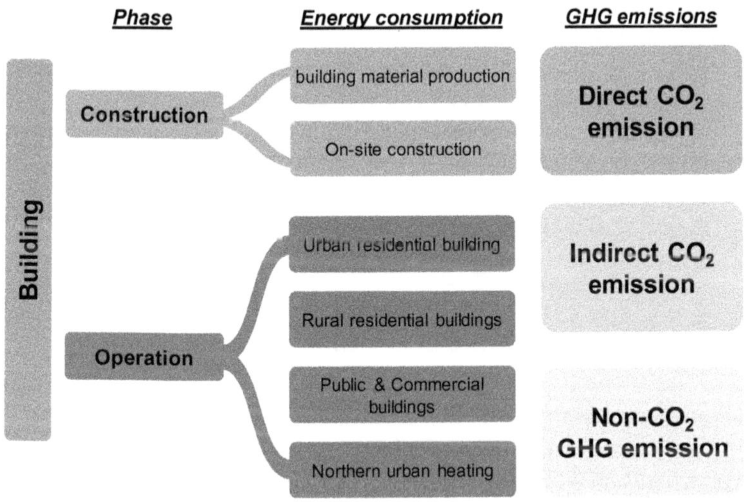

Fig. 2.6 Boundaries and types of energy consumption and GHG emissions in the building sector

The energy consumption of the building sector includes building construction energy consumption and building operation energy consumption. Building construction energy consumption refers to the energy consumption from raw material mining, building material production, transportation, and on-site construction due to building construction. In general statistical standards, civil building construction, production building (non-civil building) construction and infrastructure construction are included in the construction sector, so their energy consumption is collectively known as the construction energy consumption or emissions of the construction sector. Based on estimates of BERC, Tsinghua University, this study provides the analysis data of the standard for construction energy consumption/emissions of China's construction sector and the standard for construction energy consumption/carbon emissions of China's civil buildings.

Building operation energy consumption refers to the energy consumption from the provision of heating, ventilation, air conditioning, lighting, cooking and domestic hot water (DHW) to occupants or users in residential buildings, office buildings, schools, malls, hotels, transportation hubs, recreational and sports facilities and other buildings, and from the realization of service functions of such buildings. While studying the energy use in buildings from countries around the world, many international energy research institutions usually divide building operation energy consumption into the energy consumption of residential and non-residential buildings. However, such division cannot reflect the true types and features of building energy consumption in China. In our study, we always divide building operation energy consumption into the energy consumption by urban residential buildings, public buildings, rural residential buildings and NUH. Moreover, it is very difficult to distinguish the operation energy consumption of buildings that fully serve industrial production processes from the industrial production energy consumption, because the energy consumption in ventilation, air conditioning and purification of factory buildings, metallurgical factory buildings or integrated circuit or pharmaceutical production factory buildings accounts for a very large proportion in the production energy consumption. However, it is very difficult to include such energy consumption into building energy consumption. Hence, this report does not discuss the buildings serving production processes and only studies civil buildings.

2.2.1 *Building Operation Energy Consumption*

The building operation energy consumption in this report refers to the energy consumption in the operation of civil buildings, including residential buildings, office buildings, schools, malls, hotels, transportation hubs, recreational and sports facilities, and other nonindustrial buildings. Based on our long-term research on the energy consumption of civil building operation in China and in view of the difference in heating methods in winter between northern and southern China, the difference in architectural forms and lifestyles between urban and rural areas and the difference in personnel activities and energy use equipment between residential and public buildings, this report divides the building energy use in China into four categories, i.e. NUH energy use, urban residential building energy use (excluding NUH energy use), public building energy use (excluding NUH energy use), and rural residential building energy use, which are defined in detail as follows.

1. **NUH energy use**

This refers to the energy consumption of heating in winter, including all forms of centralized heating and decentralized heating, in provinces, autonomous regions and municipalities that adopt centralized heating, including all urban areas of Beijing, Tianjin, Hebei, Shanxi, Inner Mongolia, Liaoning, Jilin, Heilongjiang, Shandong,

Henan, Shaanxi, Gansu, Qinghai, Ningxia and Xinjiang as well as part of Sichuan. Heating is also needed in winter in Tibet, western Sichuan and part of Guizhou, but should be considered separately as the local energy situations, issues and characteristics are different from those in northern China. The reason for the separate calculation of NUH energy use is that centralized heating has been the main heating method in northern urban areas, including a large number of city-level heating networks and community-level heating networks. Different from other categories of building energy use, in which the calculation is based on the consumption of a single building or a single household, NUH energy use is largely related to the structural form and operation mode of a heating system, and the actual value of energy use is counted and calculated in a unified manner based on the heating system, so NUH energy use is taken as a separate category and treated differently from other categories of building energy use. Based on the form and scale of heat source systems, the current heating systems can be classified into centralized heating systems, which adopt such methods as large- and medium-scale combined heat and power (CHP), small-scale CHP, district coal-fired boilers, district gas-fired boilers, community-level coal-fired boilers, community-level gas-fired boilers and heat pumps, and decentralized heating systems, which adopt such methods as household gas furnaces, household coal furnaces, air conditioners and direct electrical heating. The main types of energy sources used include coal, gas and electricity. This report studies the primary energy consumption, including the primary energy consumption or electricity consumption at heat sources and the electricity consumption of equipment (fans and water pumps) serving heating systems. Such energy consumption can also be divided into the conversion loss from heat sources and heating stations, the heat loss and transmission and distribution energy consumption in pipe networks, and the final heat gains of buildings.

2. **Urban residential building energy use (excluding NUH energy use)**

This refers to the energy consumption of urban residential buildings, except for the heating energy consumption in northern China. In terms of energy end-users, it includes energy consumption of household appliances, air conditioners, lighting, cooking and domestic hot water, as well as energy consumption of heating in winter in provinces, autonomous regions and municipalities in the hot-summer and cold-winter (HSCW) zone. The main types of commercial energy sources used in urban residential buildings include electricity, coal, natural gas, liquefied petroleum gas (LPG), and city gas. Decentralized heating is mostly adopted in winter in the HSCW zone, and the energy consumption of the following heating methods all falls into this category: building space heating methods such as air source heat pumps and direct electrical heating, and local heating methods such as fire pans, electric blankets, and electrical hand warmers.

3. **Public and commercial building energy use (excluding NUH energy use)**

The public and commercial (P&C) buildings here refer to buildings where people carry out various public activities, including office buildings, commercial buildings, tourism buildings, scientific research, educational, cultural and medical buildings, communication buildings, and transportation buildings in urban and rural areas.

Except for NUH energy consumption, the energy consumption of activities in buildings includes the energy consumption of air conditioning, lighting, sockets, elevators, cooking and service facilities, as well as the energy consumption of heating of urban public buildings in winter in the HSCW zone. The types of commercial energy sources used in public buildings include electricity, gas, fuel oil, and coal.

4. **Rural residential building energy use**

This refers to the energy consumption of rural households, including cooking, heating, cooling, lighting, hot water, household appliances, etc. The main types of energy sources used in rural residential buildings include electricity, coal, LPG, gas, and biomass energy (straw and firewood). The consumption of biomass energy is not included in the national macrostatistics of energy. However, as an important part of rural residential building energy use, it will be listed separately in this report.

In this report, electricity consumption and the actual consumption of other types of energy are counted and calculated separately whenever possible. If they must be combined, all energy sources will be converted into primary energy sources for addition, namely, the electricity consumption will be converted into primary energy consumption calculated in standard coal based on the annual average coal consumption for power supply in China. As to the CHP method of centralized heating source for building operation, the input fuels are allocated based on the exergy values of output electricity and heat according to relevant provisions of the *Standard for Energy Consumption of Building* (GB/T 51161-2016).

The building energy consumption data in this chapter comes from the results of research with the China Building Energy and Emission Model (CBEEM) built by BERC, Tsinghua University, and is used to analyze the development of building energy consumption in China. From 2010 to 2020, the total energy consumption of buildings and electricity consumption increased dramatically, as shown in Fig. 2.7. In 2020, the total commercial energy consumption of building operation was 1.06 gigatonnes of coal equivalent (Gtce), accounting for about 21% of the total energy consumption in China, and the commercial energy consumption and biomass energy consumption of buildings amounted to 1.11 Gtce (biomass energy consumption: about 0.09 Gtce), with details given in Table 2.1.

The scale, intensity and total quantity of the four categories of building energy use are represented in the four blocks in Fig. 2.8 respectively, in which the horizontal axis represents building area and the vertical axis represents energy use intensity per square meter of building. The total area of the four blocks represents the total energy consumption of buildings. In terms of building area, urban residential buildings and rural residential buildings have the largest area, while NUH area and public building area account for about one-fourth and only one-fifth of the total, respectively. However, from the perspective of energy use intensity, the energy use intensity of public buildings and NUH is higher than that of the other two categories. Therefore, in terms of total energy use, each of the four categories of energy use accounts for about one-fourth of the total building energy consumption. In recent years, the growth in the scale and average energy use intensity of public buildings has made

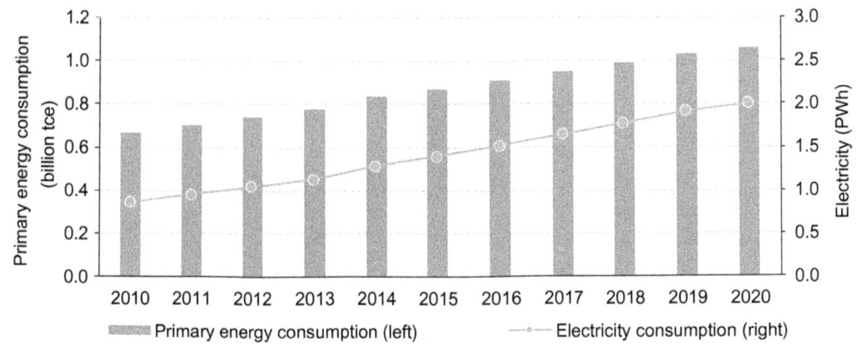

Fig. 2.7 Primary energy consumption and total electricity consumption of building operation in China (2010–2020)

Table 2.1 Building operation energy consumption in China in 2020

Energy use category	Macroparameter (billion m^2)	Electricity consumption (TWh)	Commercial energy consumption (million tce)	Primary energy use intensity
NUH	15.6	63.9	214	13.7 kgce/m^2
Urban residential building (excluding NUH)	29.2	569.4	267	759 kgce/household
Public building (excluding NUH)	14	1022.1	346	24.7 kgce/m^2
Rural residential building	22.7	344.6	229	1,212 kgce/household
Total	1.41 billion people 66 billion m^2	2000	1060	

their energy consumption the largest proportion in the building energy consumption in China.

Figure 2.9 shows the changes in the total quantity and intensity of the four categories of energy use from 2010 to 2020. The total quantity in each category of energy use showed an obvious growth, except for the continuous decline in the rural biomass energy consumption. The following characteristics are further found in the analysis of energy use intensity of buildings:

- Although NUH energy use intensity is relatively large, it has been decreasing in recent years, revealing the efficacy of energy saving and carbon reduction.
- Energy use intensity per unit area of P&C building continued to increase. The increasing terminal energy demand of public buildings (air conditioners, devices, lighting, etc.) was the major cause for the increase in building energy use intensity. In particular, some new large buildings with large-scale centralized systems have

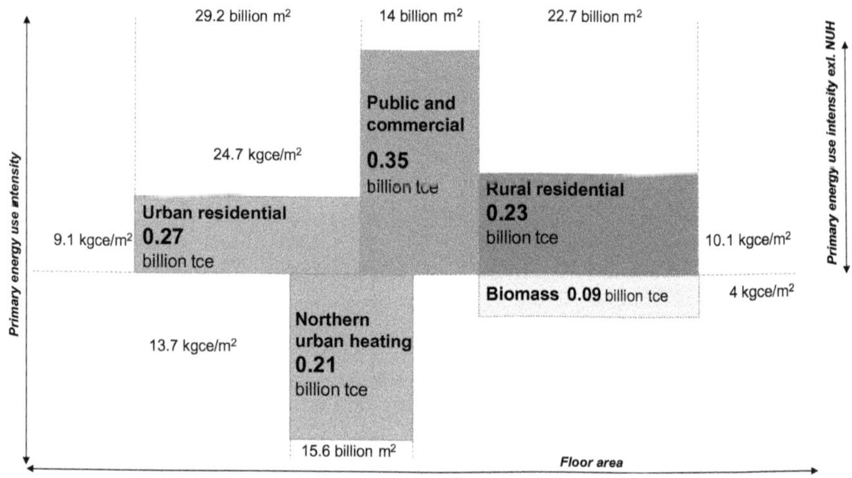

Fig. 2.8 Building operation energy consumption in China (2020)

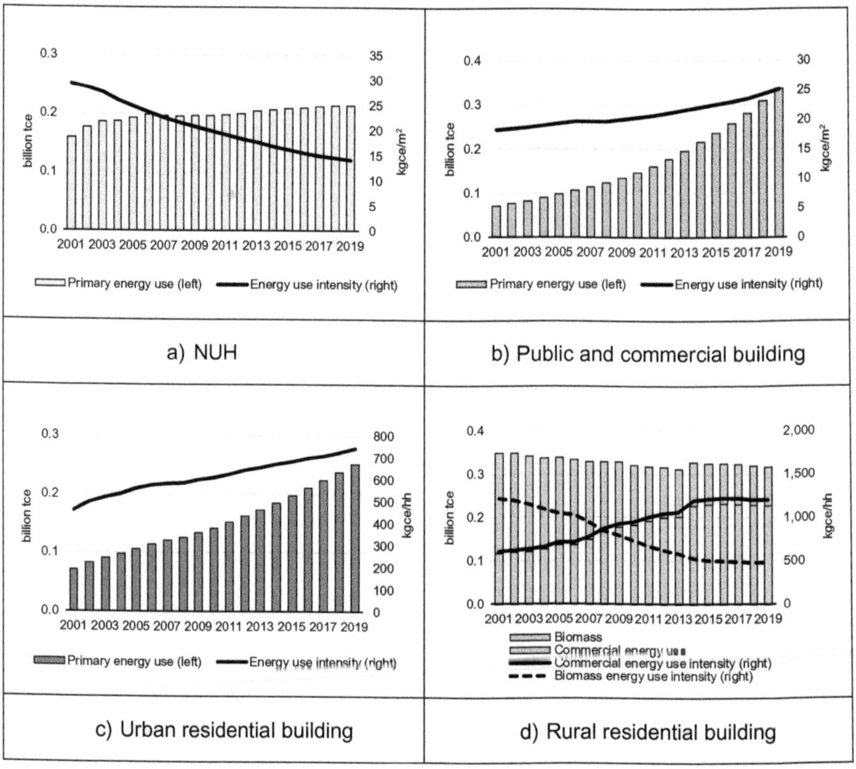

Fig. 2.9 Changes in total building energy use and energy use intensity by year (2010–2020)

been constructed in many cities in recent years, with energy use intensity much higher than that of similar buildings.

- The energy use intensity per household of urban residential buildings increased because there was an increasing demand for domestic hot water, air conditioners and household appliances. The issue of heating in winter in the HSCW zone also aroused extensive discussions. There was not too much increase in the energy consumption of lighting in residential buildings because of the adoption of energy-efficient illumination devices. The energy use intensity of cooking also remained basically unchanged.
- The commercial energy use intensity per household of rural residential buildings increased slowly. As the rural population and the number of households slowly decreased, the commercial energy consumption in rural areas basically remained stable. However, as household appliances became more popular in rural areas and the policy of "replacing coal with electricity" was implemented for clean heating in northern China, electricity consumption has increased dramatically in recent years. Meanwhile, biomass energy use has dropped continuously. Hence, the total energy use of rural residential buildings has declined slowly in recent years.

2.2.1.1 NUH

In 2020, the energy consumption of NUH was 214 million tce, making up 20% of the total energy consumption of buildings in China. From 2001 to 2020, the NUH area tripled from 5 to 15.6 billion m^2, while the total energy consumption increased, but did not double. Obviously, the increase in total energy consumption was less than the increase in building area, indicating that remarkable results had been achieved in energy saving. The average energy consumption per unit area of heating was 13.7 kgce/m^2 in 2020, a significant decline from 23 kgce/m^2 in 2001. Specifically, the main reasons for the decrease in energy use intensity include the improvement of building insulation, the increased share of efficient heat sources, and the improvement of operation management.

- Improvement of building envelope performance. In recent years, the Ministry of Housing and Urban-Rural Development of the People's Republic of China has adopted various methods to improve building insulation, including establishment of the building energy efficiency design codes that covers different climate zones and building types, the special examination of energy saving work started from 2004, and the renovation of existing residential buildings during the 13th Five-Year Plan period. These methods have greatly enhanced building insulation in China and lowered the actual heating demand of buildings especially in northern China.
- Rapidly increased share of efficient heat sources. The efficiency of heating methods varies. Overall, high energy-efficiency heating methods such as CHP for centralized heating and district boilers have been replacing small coal-fired boilers and household coal furnaces, thus leading to a rapid decrease in the shares of the

latter two heating methods. Various heat pumps have mushroomed, and the proportion of gas heating has increased. Moreover, heating system efficiency has also been increasing notably in recent years, thus enabling the overall improvement for the efficiency of all types of centralized heating systems.

2.2.1.2 Urban Residential Buildings (Excluding NUH)

Urban residential building energy consumption (excluding NUH) in 2020 was 267 million tce, accounting for one-fourth of the total commercial energy consumption in the building sector. Electricity consumption was 569.4 TWh. With the economic and social development and the improvement of living standards in China, the average annual growth rate of urban residential building energy consumption reached up to 8% from 2001 to 2020, and the terminal electricity consumption in 2020 quadrupled that in 2001.

From the view of energy use, cooking, household appliances and lighting consumed the most energy in urban residential buildings (excluding NUH) in China. Thanks to policies and projects for improving the energy efficiency of cooking, household appliances and lighting, the terminal energy consumption of these three categories was kept under control, and the total energy consumption has undergone a slower increase in recent years. Improving energy efficiency and lowering standby energy consumption should become the optimal methods to limit the energy consumption of cooking, household appliances and lighting. For example, the promotion of energy saving lamps significantly improved the lighting efficiency of residential buildings. Energy efficiency standards and behaviors need to be upgraded to lower the energy wasted from long standby time and frequent reheating and restarting of household appliances. The production standards of such appliances as televisions, water dispensers and electric toilet seats need to be improved to lower the waste of energy when they are in standby mode. The methods, for example, include improving the controllability of set-top boxes, enhancing the insulation capacity of water dispensers, and adopting intelligent control of toilet seats. Policy incentives or subsidies shall not cover electric appliances such as clothes dryers, which may result in changes in lifestyle, and we should watch for the energy spikes of these high energy consumption appliances. Even though the energy consumption of winter heating, summer cooling and domestic hot water accounts for a smaller proportion in the HSCW zone and the energy consumption per household is at a low level, they have been growing rapidly. The annual average growth rate of heating energy consumption in the HSCW zone could be well over 50%. Therefore, saving terminal energy use for those three categories should be our priority in the next stage of energy saving for urban residential buildings. We should avoid the massive adoption of centralized systems, promote decentralized systems in residential buildings, improve the energy efficiency standards of distributed equipment, and prevent against drastic energy consumption increases while improving the indoor service level.

2.2.1.3 Public Buildings (Excluding NUH)

In 2020, the public and commercial building area in China was approximately 14 billion m^2, and the total energy consumption of public buildings (excluding NUH) was 0.346 Gtce, making up 33% of that of the building sector. Electricity consumption stood at 1022.1 TWh. The total area of public buildings and the proportion of large public buildings were all on the rise, which led to an increase in energy demand. The energy consumption per unit area of public buildings grew from 17 kgce/m^2 in 2001 to over 24.7 kgce/m^2 in 2020. The energy use intensity increased rapidly and the total energy consumption surged as well.

In 2020, due to the impact of COVID-19, the operation duration and intensity of public and commercial buildings were subject to pandemic-related control measures, leading to the first slight decrease in energy use intensity in the recent decade. However, in the long run, the total scale and energy use intensity of public and commercial buildings in China are still increasing slowly. Since 2001, the as-built area of public buildings has amounted to almost 8 billion m^2, accounting for approximately 79% of the current stock of public buildings, which means that three-fourths of public and commercial buildings have been newly built after 2001. There are two reasons for this increase. First, a lot of new commercial buildings such as office buildings and commercial complexes have been constructed in recent years. Second, the scale of public service buildings such as schools, hospitals and sports stadiums has increased due to the necessity for the gradual perfection of relevant infrastructure to promote the building of a well-off society in an all-around way and the improvement of public services. Section 3.1 of this report for details about the trend of growth in public and commercial building area in recent years.

The rapid increase of public and commercial building area is also accompanied by a remarkable rise in the proportion of large commercial buildings. In particular, many public and commercial buildings completed in recent years are large top-grade commercial buildings with central air-conditioning. Their electricity consumption per unit area is over 100 kWh/m^2, while the electricity consumption of smaller schools, offices and stores built in the past is approximately 60 kWh/m^2. The average electricity consumption of public and commercial buildings will continue to increase as the proportion of such new public and commercial buildings with high energy consumption in the total number of public and commercial buildings has been on the rise. Due to the volume and form constraints of such new buildings, the energy use intensity of air conditioning, ventilation, lighting and elevators in them is much higher than that in general public and commercial buildings. This is also an important cause to the continuous growth in the energy use intensity of public and commercial buildings in China.

2.2.1.4 Rural Residential Buildings

In 2020, the commercial energy consumption of rural residential buildings was 0.229 Gtce, accounting for 22% of the total energy consumption of buildings in China.

Electricity consumption was 344.6 TWh, and the consumption of rural biomass energy (straw and firewood) was equivalent to about 0.09 Gtce. From 2001 to 2020, urbanization led to a decline of the rural population from 800 million to 510 million, and the scale of rural residential buildings was maintained at approximately 23 billion m^2, and has actually begun to decrease slowly in recent years.

Owing to the higher availability of electricity in rural areas, higher income for rural residents and more household appliances in recent years, the electricity consumption per household in rural areas has increased rapidly. For instance, the average number of air conditioners per hundred households in rural areas increased from 16 in 2001 to 71 in 2019, which led to not only the growth of electricity consumption but also longer peak power load in rural areas during summer. The implementation and promotion of "replacing coal with electricity" in northern China contributed to significant growth in winter heating electricity consumption and peak power load there. Refer to the *2020 Annual Report on China Building Energy Efficiency* for details. Moreover, increasing biomass energy has been replaced by scattered coal and other commercial energy sources, leading to rapid reduction of the proportion of biomass energy in household energy consumption in rural areas.

An important technical measure for carbon emission reduction, the utilization of biomass and renewable energy will play a huge part in rural residential buildings. The *Innovation Action Plan of Energy Technological Revolution (2016–2030)* states that eco-energy farms, biomass energy and energy crops will be developed in rural areas. *The 13th Five-Year Plan for Biomass Energy Development* clearly delineated the development goal of biomass energy use in China, that is, "promoting the application of biomass briquette in cooking and heating in rural areas" and developing biomass energy into a new industry for rural economic development. Meanwhile, China released the *Work Plan for Implementation of PV Poverty Alleviation Project* in 2014, which proposed the development of PV (photovoltaic) industry in rural areas as an important means of poverty alleviation. How to raise the living standard in rural areas without increasing commercial energy consumption and to increase the utilization of non-commercial energy by taking advantage of abundant renewable resources in rural areas and through total energy solutions is the key to energy saving of rural residential buildings, and is also an important issue for the sustainable development of China's energy system.

In recent years, with the thorough implementation of haze governance and clean heating in eastern China, governments at all levels and relevant enterprises have made huge investments to increase the power supply capacity in rural areas, lay gas pipe networks and change original small household coal-fired furnaces into low-pollution forms, which leads to a substantial increase in electricity and gas consumption. Refer to the *2020 Annual Report on China Building Energy Efficiency* for detailed data about electricity and natural gas consumption in rural areas. The change in rural energy structure will lead to a fundamental transformation of rural energy use patterns, thus facilitating the modernization of rural areas. This opportunity should be leveraged and scientific planning should be made to revolutionize rural energy supply and consumption and establish a new energy system with renewable energy sources as the mainstay for rural residents, which will play an important role in the current energy revolution of China.

2.2.2 Embodied Energy of Building Sector

As urbanization continues, energy consumption of civil building construction in China also increases. A large quantity of building materials is needed, the production process of which requires a great deal of energy and carbon emissions. This is one of the key reasons for the continuing growth of energy consumption and carbon emissions in China.

According to the estimation from the Building Energy Research Centre (BERC) of Tsinghua, embodied energy use of civil buildings in China amounted to 0.52 gigatonnes of coal equivalent (Gtce), accounting for 10% of China's total energy consumption. The embodied energy use of civil buildings in China grew from 0.24 Gtce in 2004 to 0.57 Gtce in 2016, and then gradually fell to 0.52 Gtce in 2020, as shown in Fig. 2.10. Due to the slow decrease in the newly built building stock of civil buildings in recent years, its embodied energy use has also dropped gradually since 2016. The embodied energy use of urban residential, rural residential and P&C buildings accounted for 71%, 6% and 23% in 2020, respectively.

In fact, the construction sector consists of not only civil buildings but also buildings for production purposes and infrastructures such as motorways, railways, and dams. The embodied energy use of the construction sector mainly includes all types of energy use related to the construction of buildings and infrastructures. According to the calculation of BERC,[1] Tsinghua University, total embodied energy use of China's construction sector in 2020 was 1.35 Gtce, accounting for up to 27% of the primary energy consumption of the whole society. From 2004 to 2020, embodied energy use of China's construction sector grew from approximately 0.4 to 1.35 Gtce, as shown in Fig. 2.11. Embodied energy from building materials is the mainstay of the total

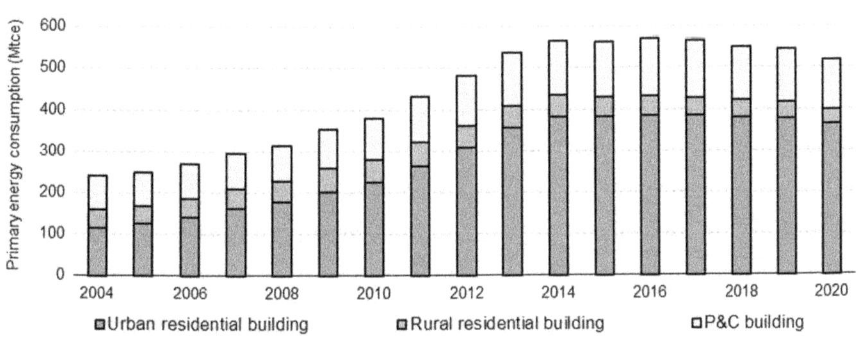

Fig. 2.10 Embodied energy use of China's civil buildings (2004–2020). *Source* Estimation by BERC, Tsinghua University. Civil building construction is included only[1]

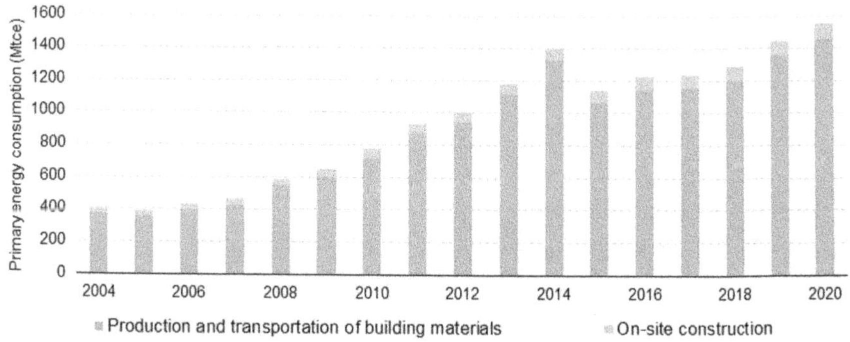

Fig. 2.11 Embodied energy consumption of China's construction sector (2004–2020).[2] *Source* Estimation by BERC, Tsinghua University. The construction sector involves the construction of civil buildings, production buildings and infrastructures

embodied energy use of buildings, in which iron and steel and cement production consume more than 70%.

The construction demands of rapid urbanization in China have not only driven the growth of energy consumption directly, but also determined China's industrial structure which is dominated by traditional heavy and chemical industries including steel and cement. This is also a key reason for the high energy consumption per unit of industrial value added in China. In 2017, the energy consumption per unit of value added in manufacturing was 6.4 tce per ten thousand yuan (constant USD price of 2010) in China, but lower than 2 tce per ten thousand yuan (constant USD price of 2010) in France, Germany, Japan and the UK. In the US and South Korea, it was 3.1 tce per ten thousand yuan and 4.5 tce per ten thousand yuan (constant USD price of 2010), respectively, which were still lower than that in China (Fig. 2.12).

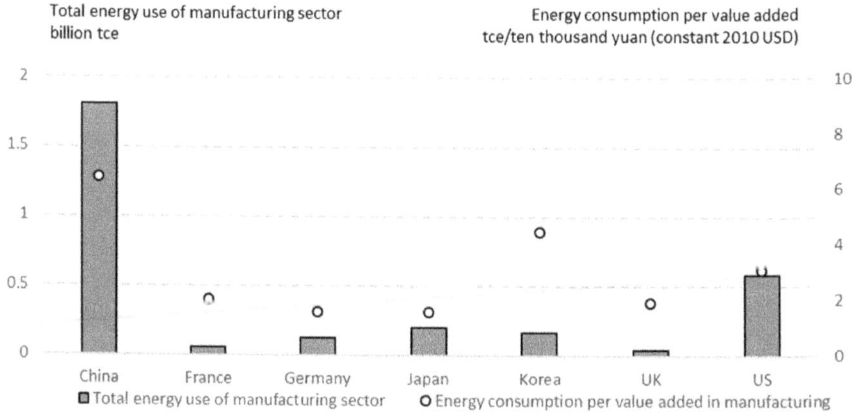

Fig. 2.12 Comparison of total energy use and energy consumption per unit of value added in manufacturing[1]

The comparison of industry structure for energy use in the manufacturing sector among countries is shown in Fig. 2.13. In 2017, the total energy use of steel, non-ferrous and building material sectors accounted for 54% of the total energy use of the manufacturing sector in China, 38% in Japan, which is relatively high, approximately 27% in France, Germany and South Korea, which was only half of China's level, and 18 and 11% in the UK and the US, respectively, which were both less than 1/3 of China's level.

According to the comparison of energy use per unit of value added among manufacturing sub-sectors as shown in Fig. 2.14, the energy consumption per unit of value added of such traditional heavy industries as steel, non-ferrous and building materials is much higher than that of electromechanical equipment manufacturing (including general machinery manufacturing, special equipment manufacturing, automobile making, and computer and communication equipment manufacturing) and also obviously higher than that of light and food industries.

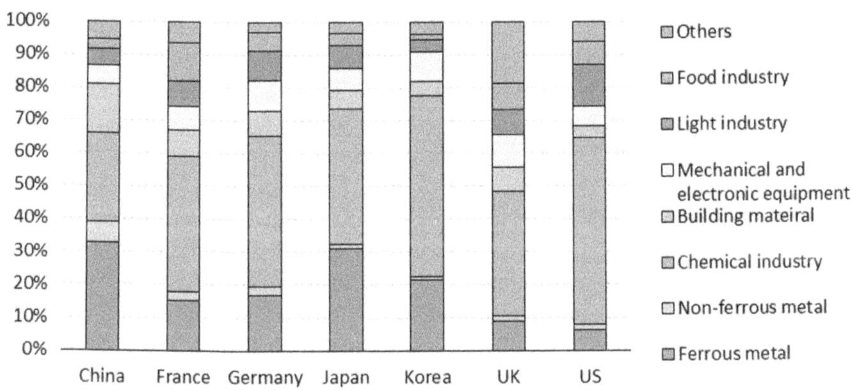

Fig. 2.13 Comparison of industry structure for energy use in the manufacturing sector between China and some developed countries[1]

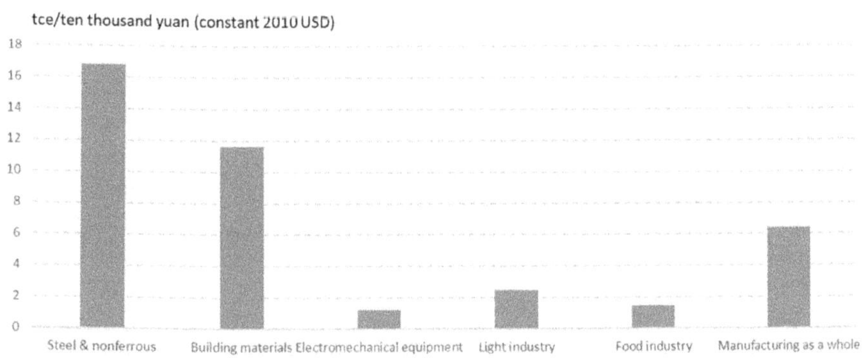

Fig. 2.14 Comparison of energy consumption per unit of value added among manufacturing sub-sectors in China (2017)[1]

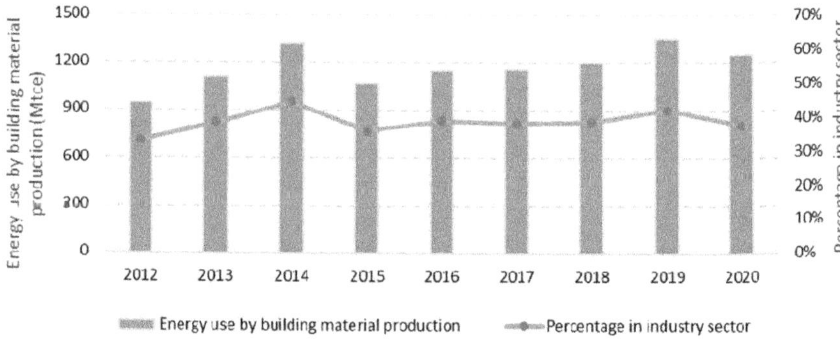

Fig. 2.15 Energy consumption for building material production[1]

Large-scale constructions have been the key foundation for China's industry structure today. In 2020, the industrial energy use by the manufacturing of materials for the construction sector was about 1.4 Gtce. Between 2013 and 2020, building materials accounted for approximately 40% of the total industrial energy consumption (Fig. 2.15). The fast pace of urbanization in China drove the demand for building materials, which was the key reason why the higher percentage of total energy consumption was represented by iron and steel, building materials and other traditional heavy industries.

As urbanization and infrastructure development have achieved initial progress in China, the transformation of the construction mode is ongoing. In 2020, the per capita floor area for urban residential buildings was 33 m^2 in China, which was close to that of some developed countries in Asia, such as Japan and South Korea, but still far lower than that of the US. The reason was that during China's urbanization, the main type of building in the urban communities was apartments instead of single houses, such as that of the US. From the perspective of urban form, the utilization ratio of P&C buildings was high in China because of the high-density large city mode of development, so it is unnecessary to follow the per capita P&C building scale in Europe and America. In the future, there will be no more rapid growth of iron and steel, building materials and other high energy consumption industries, so long as there is less demolition and the building life cycle can be properly maintained. Therefore, for the next round of urbanization, the demolition of buildings that have not reached the end of life shall be abolished. Technologies to extend the building life cycle shall be invented. Buildings and infrastructures should be properly repaired, and the building life cycle should be extended to facilitate industry transformation and total energy control.

2.3 GHG Emissions of China's Building Sector

In 2020, the central government declared that China would aim to have CO_2 emissions peak before 2030 and achieve carbon neutrality before 2060. The *Action Plan for Carbon Dioxide Peaking Before 2030* issued by the State Council in 2020 states clearly that the action for peaking carbon dioxide emissions in urban-rural development will be implemented. Therefore, scientific demarcation and quantitative analysis need to be performed on the carbon emissions in urban-rural development to drive the selection of technological paths for carbon peaking and neutrality in the building sector and for the development of work plans.

In this study, GHG emissions in the building sector are classified into categories described below. The China Building Energy and Emission Model (CBEEM) built by BERC, Tsinghua University, is used to calculate emissions of China's building sector, and analyze and discuss the present situation and trend of emissions, goals of carbon peaking and neutrality, key technologies for emission reduction, and emission reduction paths and policy measures.

(1) Carbon emission during building operation mainly includes:
Direct carbon emission: It refers to the direct emission of carbon dioxide in buildings by burning fossil fuels including coal, fuel oil and gas.
Indirect carbon emission from electricity use: It refers to the carbon emission during the generation of electricity transmitted into buildings from the outside.
Indirect carbon emission from heating: It refers to the indirect carbon emission resulting from centralized heating in northern urban areas. The centralized heating systems in northern urban areas adopt combined heat and power generation or centralized coal- and gas-fired boilers for the supply of heat. In this regard, carbon dioxide emitted by coal- and gas-fired boilers totally fall into the indirect carbon emission from building heating, while carbon emissions of combined heat and power generation plants are allocated according to the exergy values of output electricity and heat (refer to the *Standard for Energy Consumption of Building* (GB/T 51161-2016)), and the carbon emissions from heating fall into the indirect carbon emission from building heating.
(2) Indirect carbon emission resulting from building construction and repair: It refers to the indirect carbon emission due to the production and transportation of building materials and the construction of buildings during the construction, repair, and demolition of civil buildings. This part of carbon emission is also caused by construction activities of the construction sector, and belongs to the carbon emission of the building sector, but it is generally categorized as industrial carbon emission in statistics.
(3) Non-CO_2 GHG emission during building operation: It refers to the GHG effect (except for CO_2) caused by the leakage of refrigerants from refrigeration and heat pump equipment in the building sector, and is measured in carbon dioxide equivalent.

2.3.1 *Carbon Dioxide Emissions from Building Operation Energy Consumption*

The carbon dioxide emissions from building operation are affected by the growth in total energy demand of buildings, the improvement of building energy efficiency, and the adjustment of building fuel types and energy supply structure. The fuel consumed in the building operation phase are electricity, coal and natural gas. Electricity accounts for 70% of the total energy use in urban residential buildings and P&C buildings, which mainly leads to indirect CO_2 emissions. In addition, the consumption of heat from CHP for northern urban heating lead to indirect CO_2 emissions. The proportion of coal consumption was higher than that of electricity consumption in the energy consumption of NUH and rural residential buildings. The proportion of coal consumption is over 80% in NUH and about 60% in rural residential buildings, which leads to a lot of direct carbon dioxide emissions. With the increase in the proportion of zero-carbon electricity in the electric power structure of China, the average emission factor[1] of electricity in China declined tremendously to 565 gCO_2/kWh in 2020. Besides, the share of electricity consumption in building operation energy consumption gradually increased as well. These two trends have promoted the low-carbon development of building operation energy consumption.

According to the analysis results from CBEEM, in 2020, the total carbon emissions during building operation in China were 2.18 billion tCO_2, equivalent to 1.5 t per capita carbon emissions and 33 kg/m^2 of average carbon emissions per unit area. In the total carbon emissions, the proportions of direct carbon emissions, indirect carbon emissions from electricity use, and indirect carbon emissions from heating were 27%, 52% and 21%, respectively, as shown in Fig. 2.16.

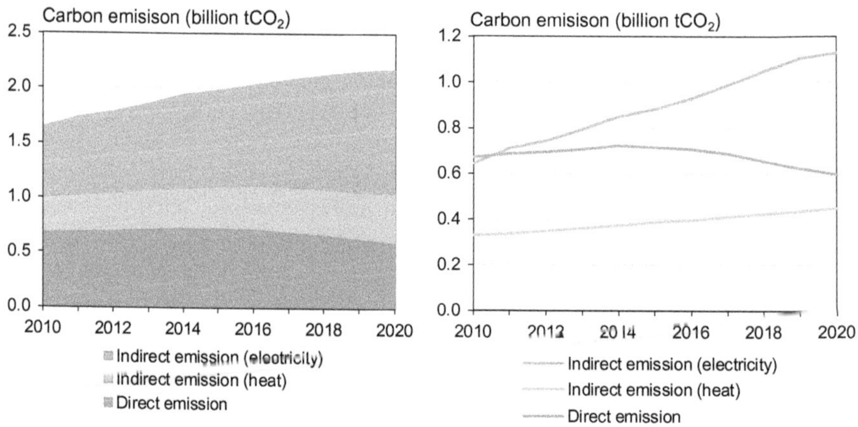

Fig. 2.16 Carbon dioxide emissions from building operation (2020)

2.3.1.1 Direct Carbon Emission

In 2020, the direct carbon emissions of buildings were 600 million tons of CO_2, including about 200 million tons of CO_2 directly emitted from urban and rural cooking, about 300 million tons of CO_2 emitted from household gas- and coal-fired heating boiler[1] and 100 million tons of CO_2 directly emitted by natural gas consumption for hot water, steam boilers and absorption refrigeration. Emissions from rural areas accounted for more than half of the 600 million tons of direct carbonemissions.

In recent years, with the vigorous promotion of "replacing coal with electricity" in rural areas, the direct carbon emissions in the building sector of China have been declining slowly from the peak around 2015. Direct carbon emissions of the building sector will continue to decrease and no longer reach a new peak as long as electrification is continuously promoted in new buildings.

To reduce direct carbon emissions of the building sector to zero, the key lies in the point-in-time and intensity of promoting "electrification". It is expected that direct carbon emissions of buildings will be reduced to zero during 2040–2045. Analyses indicate that electrification will not increase the operation cost in 80% of all cases and that the initial investment in equipment will be returned in about 5 years. Therefore, the major obstacle of promoting electrification in buildings is not economic cost but the change in the concept of energy use and cooking culture. Increasing the publicity of "zero carbon emissions of buildings by electrification" among the public and promoting "replacing gas with electricity" in new and existing buildings are the most important approaches to realize zero direct carbon emissions from building operation.

2.3.1.2 Indirect Carbon Emission from Electricity Use

In 2020, the electricity consumption by building operation in China was 2 PWh, and the indirect carbon emission from electricity use was 1.13 billion tons of CO_2. At present, the per capita electricity consumption of the building sector in China is one-sixth of that in the US and Canada and about one-third of that in France and Japan, while the electricity consumption per unit area of buildings in China is one-third of that in the US and Canada. The difference in lifestyle and building operation is one of the main reasons for the difference in electricity use intensity between China and developed countries.

In recent years, the carbon emission increase caused by the growth in electricity consumption of buildings has exceeded the carbon emission decrease resulting from the reduction of carbon emission factor of electricity. The indirect carbon emissions from electricity consumption of buildings will keep increasing before peaking. China should maintain a green and economical lifestyle and building operation mode to avoid any surge in energy use of buildings that once occurred after rapid economic growth in the history of the US, Japan, and other developed countries. When the building area in China reaches 75 billion m^2 in 2060, the electricity consumption of buildings should be 3.8 PWh, which will meet the demand of the Chinese people for good life and energy consumption of buildings. On this basis, the new type of electric power system adopting "photovoltaic, energy storage, direct current and flexibility (PEDF)" technologies should be promoted. When the reduction in indirect carbon emissions from building electricity use caused by the increase of "green electricity" and "PEDF" buildings is greater than the growth in indirect carbon emissions from building electricity use caused by the growth in the total scale and electricity use intensity of buildings each year, the indirect carbon emissions from electricity use of buildings in China can peak. Upon overall popularization of the "PEDF" power distribution mode, the zero-carbon goal for electricity consumption of buildings can be achieved earlier than that for the national power system.

2.3.1.3 Indirect Carbon Emission from Heating

In 2020, the building area of NUH in China was 15.6 billion m^2, and the indirect carbon emission from heating during building operation was 450 million tons of CO_2. Recent years have seen continuous growth in the centralized heating area and heating demand in northern China but a continuous decrease in the energy consumption and carbon emissions per square meter of heat supply. The total indirect carbon emissions from NUH showed a trend of slow growth. The indirect carbon emissions from heating during building operation can peak by around 2025 through further strengthening the renovation of existing buildings, fully exploiting low-grade residual heat resources and phasing out scattered coal-fired boilers. Afterward, with the gradual completion of the zero-emission transformation of the remaining thermal power (to be replaced with CCUS and biomass fuel) by the electric power sector, the indirect carbon emissions from building heating can be reduced to zero in sync with decarbonization process of power system.

To this end, it is necessary to continue improving envelope performance of new buildings and the renovation of existing buildings in a strict manner, so that the average heating demand of buildings in northern China can decrease from 0.35 GJ/m^2 at present to below 0.3 GJ/m^2. During 2020–2035, the return water temperature should be lowered by the terminal renovation of centralized heat supply systems, thus effectively recovering the residual heat of thermal power plant and the industrial low-grade residual heat. The increasing building heating demand will be met through digging the potential heat supply capacity of existing heat sources. CHP transformation will be carried out for the coastal nuclear power in northern China to supply heat to areas within 200 km from the normal of coastal northern China. From 2035, in coordination with the schedule of thermal power shutdown of the electric power system, seasonal heat storage projects will be implemented simultaneously to solve the problem of reduced heat sources resulting from the shutdown of thermal power plants. Until 2045, the annual residual heat of nuclear power, the annual residual heat of peak shaving thermal power, as well as the heat emitted annually from the industrial low-grade residual heat will have been collected through seasonal heat storage projects. In this way, the indirect carbon emissions from building heating can be reduced to zero, while zero carbon emission of the power system is realized.

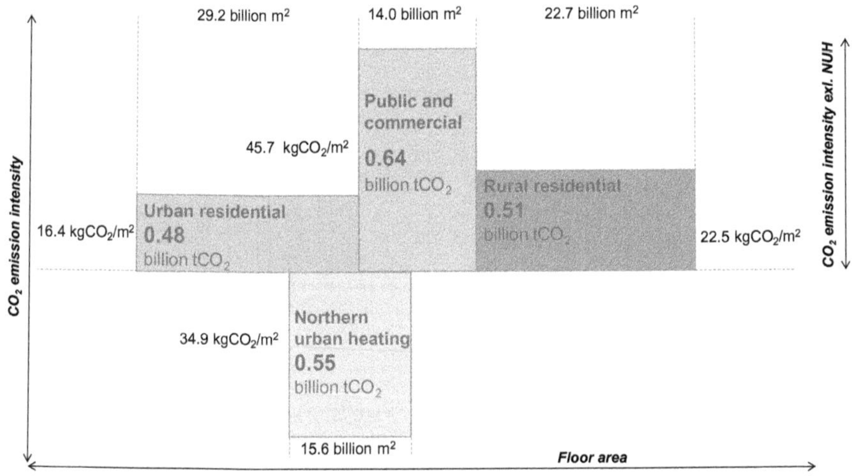

Fig. 2.17 Carbon dioxide emissions from building operation in China (2020)

In view of the four categories, the scale, intensity and total quantity of carbon emissions from them are represented in the block diagram in Fig. 2.17 respectively, in which the horizontal axis stands for the building area and the vertical axis for carbon intensity per square meter for the four categories. The total area of the four blocks represents the total carbon emissions. The growth in the total carbon emissions from the four categories is shown in Fig. 2.18. It can be seen that the characteristics of carbon emissions from the four categories are not exactly the same as those of their energy consumption. As public and commercial buildings have the highest energy use intensity, their carbon intensity per unit area is also at the peak. In 2020, their carbon intensity was 45.7 $kgCO_2/m^2$, and their total carbon emissions were still on the rise as their total energy consumption and energy use intensity were growing steadily. For NUH, its carbon intensity is only second to that of public and commercial buildings due to the consumption of a lot of coal. In 2020, the carbon intensity was 34.9 $kgCO_2/m^2$, and the carbon emissions basically peaked and remained stable at approximately 550 million tCO_2 due to the increasing heat demand, the improvement of heat supply efficiency and the basically consistent speed in the transformation of energy structure. Although there is little difference in primary energy use intensity per square meter between rural and urban residential buildings, the carbon intensity per square meter of rural residential buildings is higher than that of urban residential buildings due to a low electrification level and a high proportion of coal fuel. The carbon intensity per unit area of rural residential buildings was 22.5 $kgCO_2/m^2$. Due to the implementation of "replacing coal with electricity" and "replacing coal with gas" in rural areas, the total carbon emissions of rural residential buildings have peaked and then declined year by year in recent years. The carbon intensity per unit area of urban residential buildings was 16.4 $kgCO_2/m^2$, and was increasing slowly with the growth in electricity consumption.

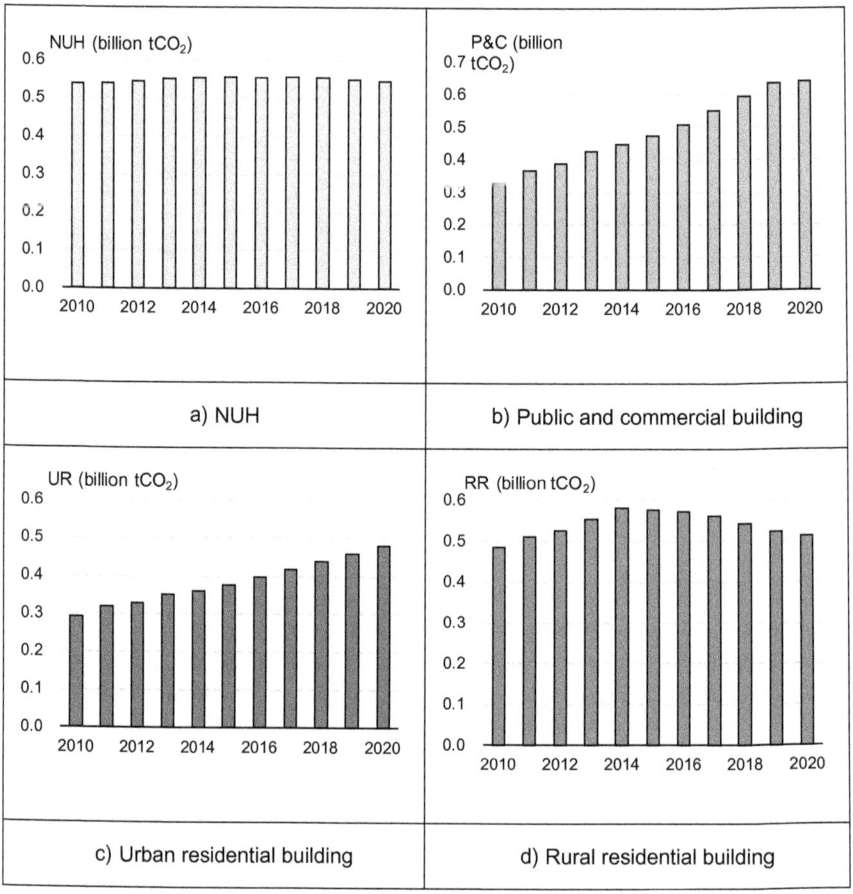

Fig. 2.18 Carbon emissions of building energy use (2010–2020)

2.3.2 *Embodied CO_2 Emission of Building Sector*

As urbanization continues, the construction energy consumption of civil buildings in China also increases rapidly. Large-scale construction activities require a large number of building materials, the production of which results in a great deal of energy consumption and carbon emissions. This is a key reason for the continuing growth of energy consumption and carbon emissions in China. The construction of buildings and infrastructures not only consumes a colossal amount of energy, but also leads to a lot of carbon dioxide emissions. In addition to carbon dioxide emissions resulting from energy consumption, emissions in the cement production process[1] are also an important part.

In 2020, the total carbon emissions from civil building construction in China were about 1.5 $GtCO_2$, mainly including carbon emissions from energy use for the manufacturing and transportation of building materials (77%), the industrial process emissions of cement (20%) and energy use during construction (3%), as shown in Fig. 2.19. Although such carbon emissions are included in the industry and transportation sector, they are driven by the demand of the building sector. Hence, the building sector shall also be responsible for such carbon emissions and reduce its demand to contribute to emission reduction. With the end of the large-scale construction period in China, the scale of newly built building stock has been decreasing each year. The carbon emissions from civil building construction peaked in 2016 and have been declining slowly year by year in recent years.

In fact, as China is still in the urbanization stage, various infrastructures need to be constructed in addition to civil buildings. In 2020, the total carbon emissions from construction in the construction sector were about 4 $GtCO_2$, nearly half of China's total carbon emissions, as shown in Figure 2.20. Carbon emissions from civil building construction were about 40% of the total carbon emissions from construction in China's construction sector.

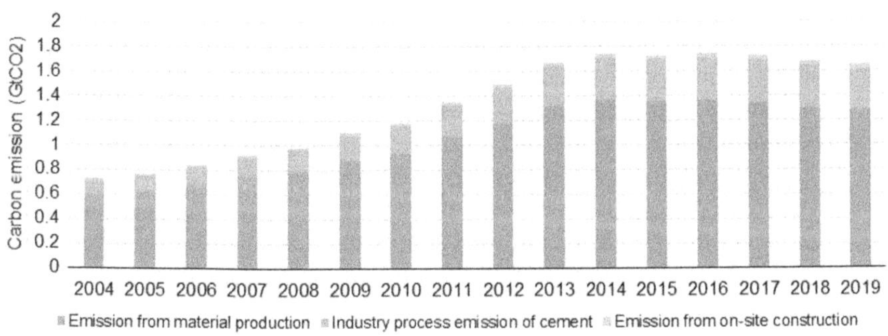

Fig. 2.19 Embodied carbon emissions of civil buildings (2004–2020).[1] *Source* Estimation by BERC, Tsinghua University. Civil building construction is included only

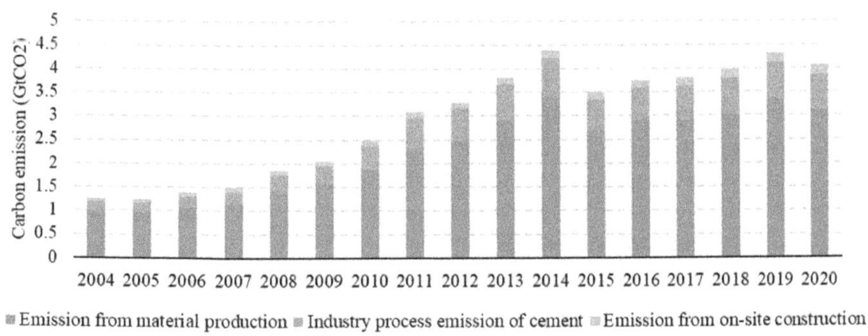

Fig. 2.20 Embodied carbon emissions of the construction sector in China (2004–2020). *Source* Estimation by BERC, Tsinghua University. The construction sector involves the construction of civil buildings, production buildings and infrastructures

To realize zero carbon emissions from building construction as early as possible, the total number and scale of buildings should be controlled reasonably first to minimize excessive construction and avoid large-scale demolition and construction. At present, the total building stock and per capita floor area in China have basically met the demands of urban and rural residence, production and living. By 2060, the production and living demands of the future urban and rural populations in China can be met with 40 m^2 of per capita residential floor area, 15.5 m^2 of per capita P&C building stock and 75 billion m^2 of total floor area. To realize zero carbon emissions from building construction in China, the construction speed, total quantity and scale of buildings need to be planned rationally.

In the meantime, China's construction sector will shift from large-scale construction of new buildings to the maintenance and functional improvement of existing ones. According to Fig. 1.2 "Newly built and demolished building stock of civil buildings in China" of this chapter, the annual newly built building stock was much more than the annual demolished building stock in the early 2000s, resulting in a net increase in the total number of buildings to meet the rigid demand for buildings. In recent years, however, the annual demolition of floor space has reached almost 2 billion m^2, although the annual total area of urban residential and public buildings completed has remained between 3 and 4 billion m^2 in those urban areas. It also shows that the construction of houses in China has shifted from increasing the supply of houses to meet immediate needs to demolishing old ones and building new ones to improve building performance and functionality. The "Mass-Demolition-for-Mass-Construction" pattern has become the main mode of the construction industry. However, according to statistics, the average lifetime of a demolished building is just over 30 years, far shorter than the lifetime designed for the building.

The main objective of the "Mass-Demolition-for-Mass-Construction" pattern is to improve the performance and functionality of buildings and to optimize land use. The big driving force behind this is the soaring land price. However, if such a demolition spree continues, it will make the construction of houses an ongoing industry rather than a periodical phenomenon. The resulting strong demand for iron and steel and building materials will then continue, as will their production. Carbon emissions thus caused and maintained will be difficult to reduce.

In fact, compared to such a mass demolition for mass construction patterns, the repair and renovation of buildings can also meet the need for functional upgrading, but a large amount of steel and cement will not be required if the main body of the structure is not involved, resulting in far less carbon emissions. Changing the mode from mass demolition for mass construction to repair and renovation can significantly reduce the consumption of building materials and thus also reduce the carbon emissions from the production of building materials. The construction industry should undergo a transformation, from building new houses to repairing old ones. Such a transformation will greatly reduce the heavy demand for building materials such as iron and steel and cement for housing construction, thereby reducing the production of such industries and transforming them accordingly.

Based on the objective of total number planning for civil buildings in the future and in view of the reasonable construction speed, the gradual transformation from "Mass-Demolition-for-Mass-Construction" to "replacing demolition with fine repair" will enable the stabilization of China's construction sector and the gradual reduction of carbon emissions from civil building construction to 200 million tons of CO_2. Zero emissions from building construction are expected to be realized in 2050 through further application of new building materials and new structure systems and technologies.

2.3.3 Non-CO_2 GHG Emissions in the Building Sector

In addition to carbon dioxide, refrigerants used in refrigeration, air conditioning and heat pump products for buildings are also GHGs causing global temperature rise. Hence, non-CO_2 GHG emissions from refrigerant leaks of refrigeration, air conditioning and heat pump products are also an important part of carbon emissions from buildings. Non-CO_2 GHG emissions in China's building sector mainly include the emissions of chlorofluorocarbon-based refrigerants from domestic air conditioners, water chiller/heater units, multi-split air conditioners and unitary air conditioners. Chlorofluorocarbon-based refrigerants commonly used in China at present include hydrochlorofluorocarbons (HCFCs) and hydrofluorocarbons (HFCs), which are mainly R22, R134a, R32 and R410A, with detailed types listed in Table 2.2. Because HFCs have the potential of zero ozone-depletion, they are regarded as the ideal replacement refrigerant for avoiding ozone depletion. However, the global warming potential (GWP) value of HFCs is relatively higher, as shown in Table 2.3, so HFCs have become the major source of non-CO_2 GHG emissions in the building sector.

Based on the estimation results from CBEEM, the refrigerant leakage caused by building air conditioning and refrigeration in China was 110 million $tCO_{2\text{-eq}}$ in

Table 2.2 HCFCs and HFCs refrigerants commonly used at present in China

Refrigeration field	HCFCs	HFCs	Others
Room air conditioner	HCFC-22	R410A, R32	
Unitary/multi-split air conditioner	HCFC-22	R410A, R32, R407C	
Water chiller unit/heat pump	HCFC-22	R410A, R134a, R407C	
Heat pump water heater	HCFC-22	R134a, R410A, R407C, R417A, R404A	CO_2
Industrial/commercial refrigeration	HCFC-22	HFC-134a, R404A, R507A	NH3, CO_2 NH_3, CO_2
Transportation air conditioning	HCFC-22	HFC-134a, R410A, R407C	
Transportation refrigeration	HCFC-22	HFC-134a, R404A, R407C	

Table 2.3 GWP values of common refrigerants

Refrigerant type	Refrigerant name	GWP values by Montreal protocol
HFCs (hydrofluorocarbons)	HFC-134a	1430
	HFC-32	675
HFCs (hydrofluorocarbon mixtures)	R-404A	3922
	R-410A	2088
	R-407C	1774
HCFCs (hydrochloroflurocarbons)	HCFC-22	1810
	HCFC-123	79

2019 and about 130 million $tCO_{2\text{-eq}}$ in 2020, which were mainly from the repair and disassembly of room air conditioners and the disassembly of commercial chillers.

The issue of non-CO_2 GHGs is as important as CO_2 in affecting climate change, and needs to be taken seriously by the building sector. In particular, with China's advance in peaking carbon dioxide emissions and achieving carbon neutrality, the proportion of non-CO_2 GHG emissions in the global total GHG emissions will gradually increase. For the building sector, non-CO_2 GHG emissions will be increasingly important to achieve climate neutrality. On September 15, 2021, the Kigali Amendment officially came into force in China. It specifies HFCs phase-down schedules and sets out that China and other developing countries in Group I shall freeze the production and consumption of controlled HFCs at the baseline level from 2024 and gradually reduce them to no more than 20% of the baseline in 2045. The further urbanization and improvement of people's living standards in China will contribute to rapid growth in the total quantity of future refrigeration equipment. This will pose a huge challenge to the non-CO_2 emission reduction in the building sector.

To reduce non-CO_2 GHG emissions from buildings, the following efforts shall be made.

(1) **Actively promoting the research and development and substitution of low-GWP refrigerants**

Refrigerant substitution has a huge influence on the refrigeration and air conditioning industry of China. Therefore, in the selection of reasonable refrigerant alternatives, it is necessary to consider the direct reduction of non-CO_2 GHG emissions resulting from refrigerant alternatives, the potential decrease in the energy efficiency of refrigerant alternatives and the resulting increase in indirect carbon dioxide emissions from electricity use. In the selection of alternative routes, various factors shall be considered to determine the GWP limits and the change point-in-time of refrigerants for different applications in China. It should be noted that most of the patents for new types of low-GWP refrigerants (HFOs) are not possessed by Chinese enterprises and that unreasonable alternative route selection may lead to the payment of considerable patent fees by the refrigeration, air conditioning and heat pump industry of China and eventually to the impairment of its competitiveness. Therefore, it is extremely urgent for China to develop low-GWP refrigerant alternatives and relevant technologies with proprietary intellectual property rights. In refrigerant substitution, priority shall be given to the refrigerants of which China holds patent rights or announces the rights.

For the development of low-GWP alternatives to HFCs and their mixtures in the field of medium- and small-capacity refrigeration, air conditioning and heat pump, natural refrigerants (HCs, ammonia, carbon dioxide, etc.) will be an important direction for future development, which is more suitable for China. Carbon dioxide is an optional natural refrigerant. Since it has a critical temperature of 31.2 °C, the operating condition of the heat pump required to release heat is a variable temperature rather than a phase transition temperature as for other types of refrigerants, which makes it possible to match heat exchange between the refrigerants and the thermal transfer media, thus increasing heat pump efficiency. Over the last two decades, heat pumps using carbon dioxide as refrigerant have been a great success. Due to the high working pressure of carbon dioxide, the required pressure-bearing capacity of compressors and systems is considerable, and China still needs to make more efforts to improve its manufacturing technology in this regard. It is therefore a critical task to address non-CO_2 GHG emissions. Cooperation across industrial borders should be encouraged to tackle the key issues and to develop self-owned packages of technology and products as soon as possible.

In the field of industrial and commercial refrigeration where refrigeration equipment can be placed separately and kept at a proper distance from people, natural refrigerants (including ammonia) that have certain safety risks but good thermodynamic properties are promising. Ammonia is a refrigerant that was used when gas compression refrigeration was first adopted. Later, it was phased out due to safety issues. For the consideration of alternatives to fluorine-based refrigerants, ammonia returns to the forefront. Through a number of innovative technologies, it is possible to overcome some of the original problems of ammonia systems, and it is likely that ammonia will gain a certain market share in the field of refrigeration, freezing, and air conditioning in the future.

However, in the field of large-scale water chiller unit, China currently does not have any refrigerant alternatives that can sidestep restrictions from other countries. At present, US companies have developed ultra-low-GWP HFO refrigerants that can be used as alternatives for a long time. China needs to develop new refrigerants and new production technologies for HFO refrigerants to get rid of technology restrictions as soon as possible. The priority shall be given to tackling the problems of the R134a alternative refrigerants for large-scale water chiller units.

(2) **Strict management of refrigerants during repair and disassembly process**

Analyses based on available data show that the leakage of refrigeration products in China now mainly occurs during repair/maintenance and final disassembly of equipment.

Leakage during operation. Refrigerants will not cause any greenhouse effect until they are released into the atmosphere. If the leak-free air conditioning and refrigeration operation can be achieved through improvement in sealing process, zero emissions during such operation can be achieved accordingly. The continuous improvement of manufacturing and installation technologies for refrigeration and air conditioning products in China has contributed to a slump in the leakage of refrigerants during operation. In particular, after the adoption of R32, R290 and other flammable or mildly flammable refrigerants in a huge number of room air conditioners, almost all refrigeration, air conditioning and heat pump products including air conditioners have been installed and maintained by specialized technicians, thus leading to the slump in the leakage of refrigerants during installation and operation. For static refrigeration, air conditioning and heat pump equipment, their pipelines can always maintain a low leakage rate due to the absence of shaking and vibration. It is estimated that the annual leakage rate of refrigerants during operation can be as low as 0.3%.

Leakage during repair/maintenance. For large-scale chillers and heat pump devices, due to the large number of refrigerants charged into them, such refrigerants are generally pumped out or stored in non-maintenance equipment during repair/maintenance, thus minimizing leakage. Nevertheless, for small-size air conditioning equipment similar to room air conditioners, once the refrigeration system fails and needs to be repaired, the refrigerants will be totally discharged into the atmosphere in most cases. It is estimated that the equivalent annual rate of leakage resulting from the repair/maintenance of household air conditioners is 0.8–1.6%.

Leakage during final disassembly of equipment. Improper treatment during equipment disassembly will lead to the emission of a lot of refrigerants into the atmosphere, so equipment disassembly is the most important process to prevent refrigerant leakage. For the moment, although clear-cut refrigerant recovery requirements have been enforced for large-scale chillers and heat pumps in China, the actual proportion of recycled refrigerants is still very low. The complete emission of refrigerants into the atmosphere remains general situation for the disassembly of small-size air conditioners.

Therefore, standardizing the repair process and recovering refrigerants during disassembly are the key. China shall establish policy mechanisms and technology systems for refrigerant recycling as soon as possible. Now, the annual recovered refrigerants are less than 1% of the annual consumption in China, while the refrigerant recovery rate in countries such as Japan is about 30%. The main reason for this is that China has not yet established a perfect refrigerant recovery system and recycling enterprises cannot obtain sufficient recovered refrigerants for production. As a result, the refrigerant recycling costs are excessive, and recycled refrigerants have no price advantage compared to newly produced ones, which in turn inhibits the will to recover and regenerate refrigerants. Therefore, strengthening regulation and establishing relevant policies, economic incentive systems and technology systems are critical to promoting the recovery, regeneration, reuse and resolving of refrigerants.

(3) **Actively promoting refrigerant-free refrigeration and heat pump technologies**

In addition, another technological path of reducing GHG effects caused by refrigerant leakage is to develop new refrigerant-free refrigeration technologies and completely avoid the use of fluorinated gases in certain occasions where leakage cannot be avoided or is difficult to manage. Currently, all countries in the world are researching and developing non-vapor-compression refrigeration and heat pump technologies. Through the use of indirect evaporative cooling technology in dry regions, cold water below the wet bulb temperature can be obtained to meet the needs for comfortable air conditioning and data center cooling, while electricity consumption by such cooling is significantly reduced. By taking advantage of low-grade heat emitted at approximately 100 °C and by means of absorption cooling, the cooling sources required for comfortable air conditioning and industrial air conditioning can be similarly obtained. As residual heat is used, the benefits of energy saving can be achieved. In addition, solid-state refrigeration technologies such as thermoacoustic refrigeration, magnetic refrigeration and semiconductor refrigeration have huge potential due to direct electric driven refrigeration without refrigerant. In recent years, solid-state refrigeration technologies have seen major theoretical and technical breakthroughs, with increased refrigeration capacity and efficiency, and are gradually penetrating into the building sector.

The issue of non-CO_2 GHGs is as important as CO_2 in affecting climate change, and needs to be taken seriously by the building sector. A solution to the problem of non-CO_2 GHG emissions, when made available, will lead to revolutionary changes in freezing, refrigeration, and air conditioning technologies in buildings and will bring about innovative breakthroughs in technologies, which deserves attention of the industry.

2.4 Energy Consumption and GHG Emissions in the Global Building Sector

2.4.1 Energy Consumption by Global Building Operation

According to the International Energy Agency's (IEA) calculation of global energy use and emissions of the building sector (as shown in Fig. 2.21), in 2020, terminal energy use during the global building construction stage (including building and infrastructure construction) and building operation accounted for 36% of the total global energy consumption, in which the terminal energy use of construction accounted for 6% and the energy consumed during the operation stage accounted for 30%. In 2020, global CO_2 emissions during the building construction stage (including building and infrastructure construction) accounted for 10% of the total, and 27% of the global CO_2 emissions came from building operation.

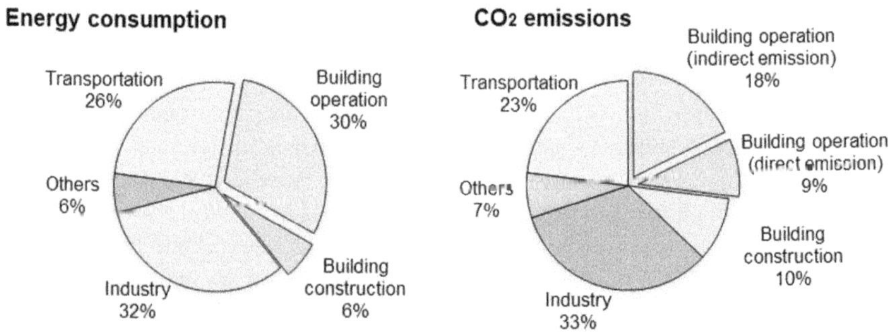

Fig. 2.21 Terminal energy use and CO_2 emissions in the global building sector (2020). *Source* 2021 Global Status Report for Buildings and Construction, International Energy Agency. The construction sector involves civil building construction, production building construction, and infrastructure construction. This figure uses the terminal energy consumption data provided by IEA, which is obtained through direct addition of heat consumption for heating, electricity consumption of buildings and terminal use of various energy types. The electricity consumption is converted to primary energy using a calorific value equivalent method. This conversion method is different from that used in the comparison of building energy consumption among countries in the ensuing part of this research, so it should be treated differently in data comparison

According to BERC's the calculation of China's building energy use and emissions, in 2020, the energy use during building construction and operation in China[1] accounted for 32% of the social total energy consumption, which was close to the global level. However, China's building construction energy use made up over 10% of the social total energy consumption, and was higher than the global level (6%). The building operation energy use accounted for 21% of the total, and was still lower than the global average. In the future, with the economic and social development of China and the improvement of living standards, the proportion of building energy use will continue to increase. Besides, in 2020, CO_2 emissions from building construction and operation in China made up about 32% of the total CO_2 emissions from energy consumption of the whole society, including 13% from building construction and 19% from building operation (Fig. 2.22).

As China's urbanization is still ongoing, the major share of energy consumption and emission has come from building and infrastructure construction. The share of energy consumption from building construction in China is higher than the global average and also higher than that in member countries of the Organization for Economic Co-operation and Development (OECD) that have already been urbanized. However, compared with those in OECD member countries, the building operation energy consumption and carbon emissions in China are at a lower level. As urbanization becomes slower in China, the proportions of building operation energy consumption and related emissions in the social total will further increase.

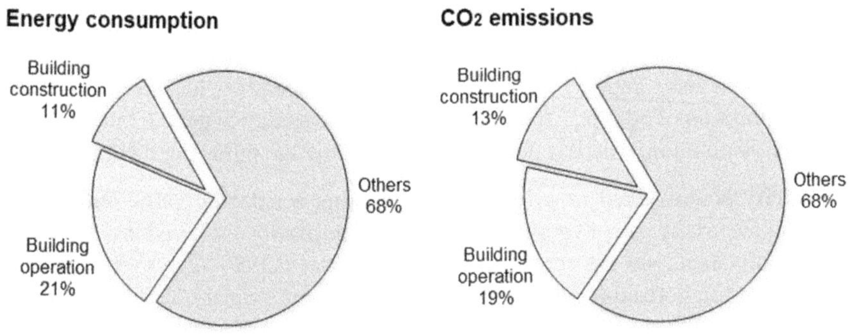

Fig. 2.22 China's building energy use and CO_2 emissions (2020). *Source* Estimation with CBEEM of BERC, Tsinghua University. The construction sector involves the construction of civil buildings, production buildings and infrastructures

2.4.2 Boundaries and Comparative Study Methods of Building Energy Consumption and Emissions

Making comparisons among countries is an important means to know the building energy consumption in China, analyze its future development trend, and design the paths of building energy saving. In this section, the building operation energy consumption and carbon emission data of countries in the world are collected and analyzed through comparison. Two types of data are collected to compare building energy consumption among countries. The first type includes population, the number of households, and building floor area. The other is building energy consumption data, which mainly includes the total consumption of electricity, heat, coal, natural gas and other fuels during building operation. The building energy data of countries around the globe collected in this study mainly comes from two sources:

(1) Databases of international organizations and agencies: mainly including IEA, Odyssee, World Bank and Eurostat databases;
(2) Official statistics of countries: For example, Japan's data mainly comes from the *Statistical Handbook of Japan 2021* and the *Japan Statistical Yearbook 2021* published by the Statistics Bureau of Japan. The data of the US is mainly sourced from the periodic surveys conducted on representative buildings of the country and the statistics released per year by the Energy Information Administration (EIA). Canada's data mainly comes from Natural Resources Canada. The data of South Korea mainly comes from the building information statistics of the Ministry of Land, Infrastructure and Transport and the KOSIS data. The data of India is mainly from the National Statistical Office (NSO) and the Ministry of Statistics and Programme Implementation (MoSPI);
(3) Some published research reports and literature also provide important support and reference for this study as they have studied the building energy and emissions in various countries and provided quantitative data.

2.4.2.1 1. Calculation of Building Energy Consumption

In the analysis and comparison of building energy consumption, it is necessary to add the consumption of all types of energy in buildings together to get the total building energy consumption, which is generally converted by the following methods.

Electricity is converted to primary energy using a calorific value equivalent. Building electricity use in various countries was uniformly converted into a thermal power equivalent, and the conversion coefficient was 0.1229 kgce/kWh; however, this approach ignores the quality and grade of different primary energy types. The same "amount" of electricity calculated by the electric-heat equivalent method was equivalent to the total building energy consumption. For example, 306 g of standard coal is required to supply 1 kWh of electricity in 2020 as per the national standard coal consumption for power supply. As such, the same "quantity" of electricity calculated using the calorific value equivalent method has a greater capacity to do work than other primary energy types. Thus, this method cannot scientifically evaluate the energy conversion process.

Convert electricity to primary energy using the primary energy consumption coefficients of thermal power supply in each country. The primary energy consumption coefficient of thermal power supply is the ratio of the primary energy consumption of coal, oil, and gas used for thermal power generation to the amount of electricity that thermal power supplies. The coal consumption of thermal power supply in each country mainly depends on the energy structure of power generation and the capacity of the power plant. For countries with smaller coal consumption for thermal power supply, the energy structure of thermal power generation in that country increases power generation efficiency and consumes less primary energy to provide the same amount of electricity; this does not mean that the country consumes less energy in the end-use sector. For example, the standard coal consumption for thermal power supply in China in 2020 was 306 gce/kWh, while that in Italy was 275 gce/kWh. The calculation results were affected by the thermal power supply efficiency in different countries. Therefore, the primary energy consumption coefficients of thermal power supply in different countries cannot be used to compare terminal energy consumption between countries.

Convert electricity into primary energy using the average primary energy consumption coefficients of power supply in each country. The average primary energy consumption factor of electricity supply is the ratio of the primary energy consumption of all energy sources used for electricity generation to the total electricity generation. As the proportion of renewable power in the power generation mix continues to increase, there is an increase to the proportion of renewable power sources, such as hydropower and nuclear power. This also significantly decreases in the average primary energy consumption coefficient of electricity supply. Countries with a large share of renewable energy, such as France where nuclear power accounts for approximately 70% of national electricity generation, it is meaningless to convert electricity to primary energy using the primary energy consumption coefficients of thermal power supply. However, this study focused on energy consumption in the end-use sector. For countries with a large share of nuclear and renewable electricity, the average primary energy consumption coefficients of the power supply are small; accordingly, the calculated primary energy consumption is low. This indicates relatively low consumption of fossil energy in that country, although it does not signify that actual consumption of end-use energy is low. For the purposes of this study, the average primary energy consumption coefficients of power supply in each country were not applicable for electricity conversion.

Convert electricity to primary energy using uniform primary energy consumption coefficients of thermal power supply. Differences in the energy structure of power generation, power plant capacity, and power generation efficiency of each country lead to variations in the standard coal consumption of thermal power supply. To avoid the power supply efficiency of each country impacting the cross-sectional comparison of building end-use energy consumption, an identical conversion factor was used for electricity conversion. For instance, China's primary energy consumption coefficients of thermal power supply can be used.

The comparison between China and other countries made in this section is mainly intended to study the actual total energy consumption and energy use intensity indicators used for buildings. Therefore, a unified primary energy consumption coefficient of thermal power should be adopted to convert the building energy use of different countries. For convenience, China's primary energy consumption coefficients of thermal power supply in that very year is adopted to convert electricity into standard coal. In this way, the influence of differences in the power supply structure and generation efficiency of each country, on the measurement of actual end-use energy may be excluded. To ensure data comparability and better reflect actual energy use, the energy consumption data in this study only included commodity energy.

2.4.2.2 2. Calculation of Carbon Emissions from Buildings

The data on carbon emissions from building operation of different countries in this section is sourced from IEA and the calculation results of the CBEEM model by BERC, Tsinghua University. When calculating the total carbon emissions from building operation, the direct carbon emissions, indirect carbon emissions from electricity use and indirect carbon emissions from heating in buildings are considered. When calculating indirect carbon emissions from building electricity use, the total carbon emissions from electricity generation were divided by the total electricity generation to obtain the average carbon emission factor for electricity use. A similar method was used to calculate carbon emissions from building heat use.

For carbon emissions during building operation, each country proposed a goal to reduce carbon emissions in the building sector. The technical pathways and priorities to achieve carbon neutrality in buildings differ between countries. To quantify and analyze the various problems confronted by each country in their attempt to achieve carbon neutrality in the building sector, the carbon emission factors for electricity and heat of each country were used in the calculation. Therefore, differences in the energy mix and efficiency among countries will affect the total amount and intensity of carbon emissions during building operation.

2.4.3 Global Energy Consumption and Carbon Emissions in the Building Sector

2.4.3.1 Building Operation Energy Consumption of Different Countries

Three indicators were selected to compare building energy use across countries: total amount, energy use per capita, and energy use per floor area, as illustrated in Fig. 2.23. The building energy use in this figure was calculated by adopting the primary energy consumption coefficients of thermal power in China to convert building electricity use to primary energy use. Energy use intensity per capita and energy use intensity per floor area are shown on the horizontal and vertical axes, respectively. The size of the bubbles represents the total primary energy use for building operation in specific countries. The bubble chart demonstrates that total energy use for building operation in China was similar to that in US, although the energy use intensity remained at a lower level. The energy consumption per capita and per floor area in China were far lower than those in the US, Canada, Europe, Japan, and South Korea.

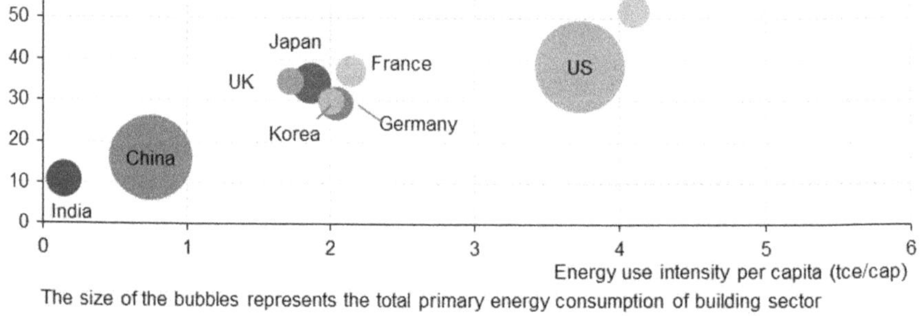

Fig. 2.23 Comparison of building operation energy consumption in different countries. Source CBEEM of BERC, Tsinghua University; World Energy Balances, Energy Efficiency Indicators database (2021 edition) of IEA; WDI database of the World Bank; Satish Kumar (2019) of India.[1] Data of 2020 is adopted for China, data of 2018 for Canada and Sweden, and data of 2019 for other countries. In the total energy consumption of buildings, the electricity consumption of different countries is converted to primary energy consumption based on China's coefficient of coal consumption for thermal power supply

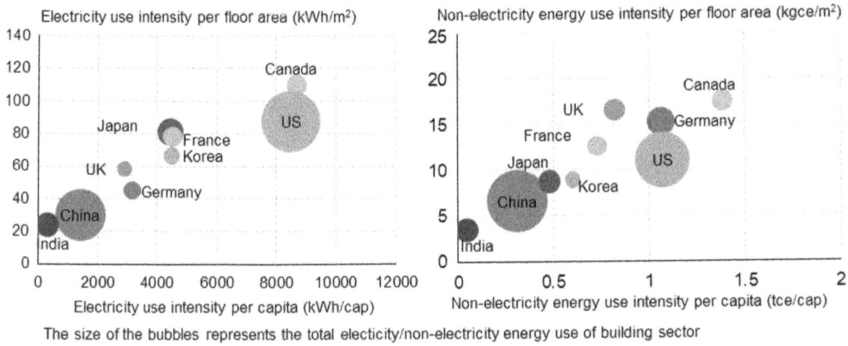

Fig. 2.24 Comparison of building operation electricity and non-electricity use in different countries. *Source* Same as above

In the context of tackling climate change and reducing carbon emissions, most countries are carrying out energy transformations, including promoting electrification in the building sector and replacing fossil fuel-based energy with renewable electricity. Figure 2.24 illustrates the electricity and non-electricity use intensities of building operation in different countries. The per capita electricity consumption in China was one-sixth that of the US and Canada; per floor area electricity consumption was one-third that of the US and Canada. However, the per capita non-electricity consumption in China was one-third of the US, and per floor area, non-electricity consumption was similar to Japan and Korea, while being half of that in the US.

China needs to develop a different route from developed countries to achieve the target for low carbon emission and energy-saving in the building sector. This would pose a significant challenge to China's low-carbon and sustainable development in the building sector. Meanwhile, there are many developing countries that are experiencing rapid changes in building energy use. China's building energy development pathway will serve as an important reference for many countries' choices, which will further influence the global building energy development.

2.4.3.2 Electrification Rate

Internationally, two indicators are often adopted to measure the degree of electrification. One is the electricity generation share of energy consumption in total primary energy use, which is used to reflect the share of electricity in the primary energy supply. The other is the share of electricity in final energy use of certain sectors, which is used to reflect the electrification rate of energy use in certain sector like building. This section compares the proportion of electricity in terminal energy consumption of the building sector among different countries. The electrification rate of energy use in buildings was calculated by converting the end-use consumption of electricity using the calorific value equivalent method.

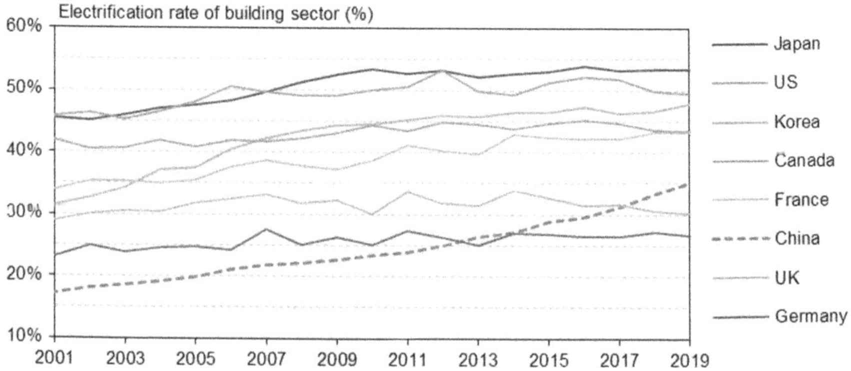

Fig. 2.25 Electrification rate of energy use in building sector in different countries (2001–2019)

Figure 2.25 compares the share of electricity in end-use energy consumption in the building sector between 2001 and 2019 among different countries. The electrification rate of the building sector in Sweden, the US, Japan, and Canada has always been high; it has exceeded 40% from 2000, and still maintains a steady growing trend. The electrification rates in France and Korea are developing rapidly, growing from ~30% in 2001 to >40% in 2019. The electrification rate of the building sector in the UK and Germany is stable and growing gradually because these two countries retain a certain percentage of fossil fuel-based energy for space heating. The electrification rate in the building sector in China has grown rapidly from 17% in 2001 to 35% in 2019, surpassing the UK and Germany, and it continues to be on a high growth phase.

2.4.3.3 Carbon Emissions of Building Sector in Different Countries

Figure 2.26 shows the total carbon emissions per capita and the proportion of carbon emissions from building sector in different countries. According to the figure, at present, the total carbon emissions per capita (including industry, building, transportation and electric power sectors) in China are slightly higher than the global average, but still lower than those in such countries as the US and Canada. The per capita carbon emissions of building operation in China are slightly higher than the global average, but obviously lower than those in developed countries. The main reason is that China is still in the process of industrialization and urbanization, and the proportion of carbon emissions from buildings in the total carbon emissions of the whole society is lower than that in developed countries. In recent years, China has been facing increasing pressure from climate change, which calls for the building sector to realize low carbon development and reach the carbon peak as soon as possible. It would remain a huge challenge for the building sector to meet the target.

Several countries have set their own goals of achieving carbon neutrality and their paths of achieving carbon neutrality in the building sector. Reducing carbon

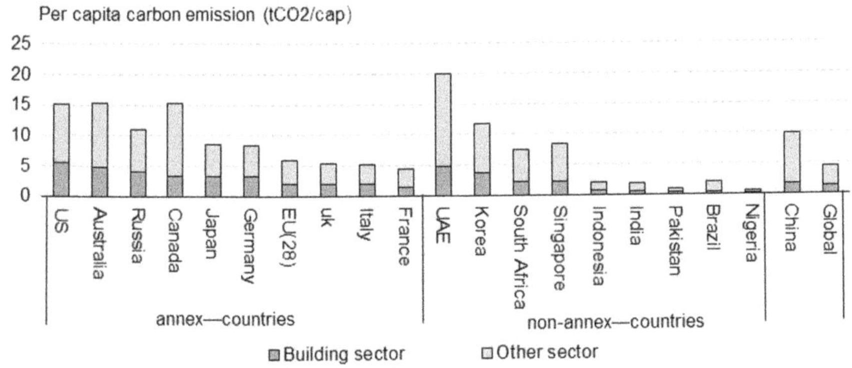

Fig. 2.26 Comparison of per capita carbon emissions of different countries (2019). *Source* Data of countries in 2019 as provided in the CO_2 Emissions from the Fuel Combustion Highlights 2020 database, IEA. Data of China are the results of 2020 as estimated with CBEEM of BERC, Tsinghua University

emissions of the building sector is also one of the important fields to realize carbon neutrality in the whole society. Figure 2.27 shows the total carbon emissions from building operation (bubble chart area), per capita carbon emissions (horizontal axis) and carbon emissions per unit building floor area (vertical axis) of different countries, which are converted according to the energy structures of these countries. The bubble chart of carbon emissions demonstrates that carbon emissions in the building sector are affected not only by the total energy consumption but also by the energy structures of these countries. The per capita carbon emissions and carbon emissions per unit floor area from building operation in China are lower than those in most developed countries due to its low building operation energy consumption. However, the energy structure of France is dominated by the low-carbon nuclear power; although it has higher building energy use intensity than China does, its carbon intensity is actually lower than that of China after conversion. This also shows that the low-carbon transition of both energy system and building energy use structure should be achieved in addition to the improvement of energy saving and energy efficiency of buildings on the path toward carbon neutrality.

In addition to comparing the current status of building energy use and carbon emissions from building operation in different countries, it is important to analyze historical trends. Figure 2.28 compares the building energy use for building operation in different countries from 2000 to 2019. The dashed circles in the figure represent carbon emissions in 2000, whereas the solid circles represent 2019 data. It reflects trends associated with total amount, intensity per capita and intensity per floor area over the past 20 years. Based on these trends, the countries can be classified into three categories. (1) The first category includes the US, Canada, Germany, the UK and France, the common characteristic of which is that their total carbon emissions, per capita carbon emissions and carbon emissions per floor area of building operation are decreasing. This benefits from the reduction in energy consumption per capita and

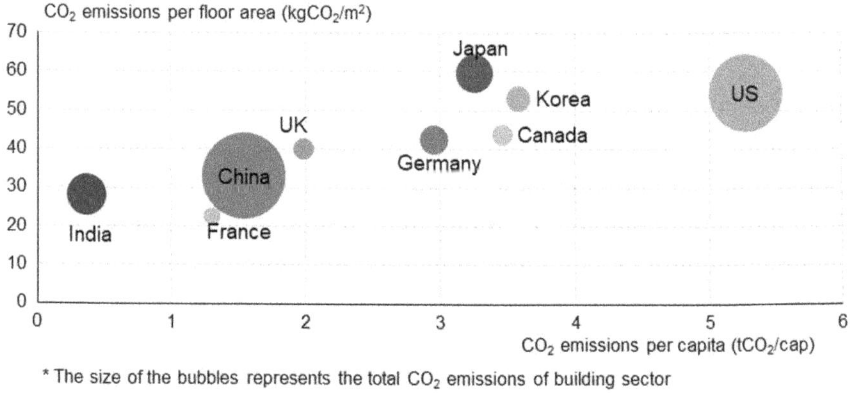

Fig. 2.27 Comparison of carbon emissions from building operation in different countries (2019). *Source* Data of countries in 2019 as provided in the CO_2 Emissions from the Fuel Combustion Highlights 2020 database, IEA. Data of China are the results of 2020 as estimated with CBEEM of BERC, Tsinghua University

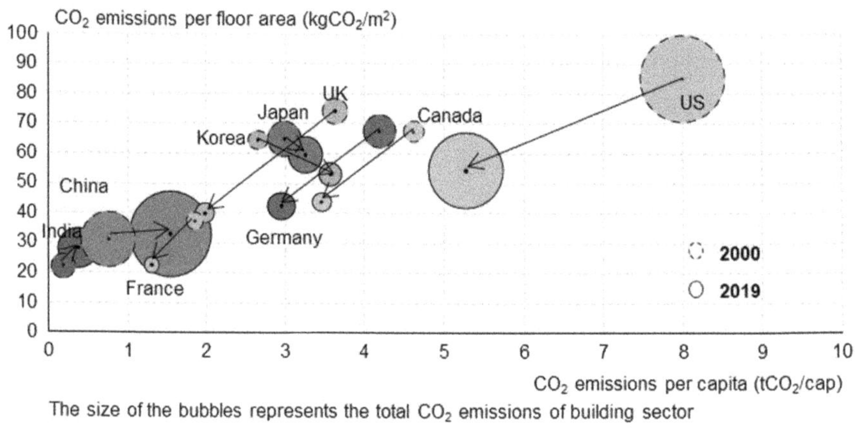

Fig. 2.28 Trajectory of building operation carbon emission in different countries (2000, 2019). *Source* Data of countries from 2000 to 2019 as provided in the CO_2 Emissions from the Fuel Combustion Highlights 2020 database, IEA. Data of China are the results of 2020 as estimated by CBEEM of BERC, Tsinghua University

per floor area and also from the active promotion of energy structure transition and vigorous development of zero-carbon electricity in these countries. (2) The second category is represented by South Korea and Japan, with increasing total carbon emissions and per capita emissions and decreasing carbon emissions per unit floor area over the last 20 years. The total carbon emissions from buildings in Japan and South Korea have experienced slow increase in recent years. Nevertheless, due to a very low growth rate of population in the last 20 years, Japan's population has shown negative growth. The growth rate of carbon emissions is greater than that of population and smaller than that of building floor area, which leads to inconsistent changes in carbon emissions per capita and carbon emissions per floor area in Korea and Japan. (3) The third category is represented by China and India, with increasing total carbon emissions, per capita carbon emissions and carbon emissions per unit area in the last 20 years. Over the past 20 years, China and India have undergone a rapid development phase, and the energy use intensity has dramatically grown. To achieve the carbon peak as early as possible, China, India, and other developing countries should not only control total energy consumption, but also urgently promote the low-carbon transition of energy systems. Improving zero-carbon renewable energy utilization in the building would be the basic principle to decarbonize building sector.

Chapter 3
Energy and Emissions of Public and Commercial Buildings

3.1 Trends in the Development of Public and Commercial Building Stock and Type

The past two decades have witnessed the rapid growth of public building stock in China from about 3.8 billion m^2 in 2001 to about 14 billion m^2 in 2020(Fig. 3.1). In 2020, office and shopping mall, as main types of public and commercial buildings at the moment, had a proportion of 34 and 23% respectively, followed by school with a proportion of 16%, while hospital and hotel had a smaller proportion of 5 and 4% respectively.

From the perspective of growth trend, the growth rate of shopping mall building stock in China has increased rapidly since 2010. The stock increased by about 0.6 billion m^2 between 2001 and 2010, and by 2 billion m^2 between 2010 and 2020, which was more than 3 times that of the first decade of the twenty-first century.

The same trend has also been reflected in the newly built building stock, see Fig. 3.2. The newly built building stock of shopping mall has shot up year by year since 2010 from less than 100 million m^2 to over 250 million m^2 per year, and exceeded that of office buildings in 2014, making shopping mall a type of public and commercial building with the largest annual newly built floor area.

Due to the change in China's urbanization process, the newly built building stock of public and commercial buildings has decreased slowly year by year since 2016 from approximately 850 million m^2 to approximately 750 million m^2 per year. Structurally, the trend of change in newly built building stock of public and commercial buildings varies with different types. The newly built building stock of office, shopping mall and hotel has been declining since 2016, while that of hospitals, schools and other types has still been on the rise.

Except for public and commercial buildings with indicated purposes mentioned above, other types of mainly include public buildings for cultural, sports and entertainment purposes and transportation hub. Such buildings fell into the category of

Y. Jiang et al., *Decarbonize Public and Commercial Buildings*,
https://doi.org/10.1007/978-981-19-5526-6_3

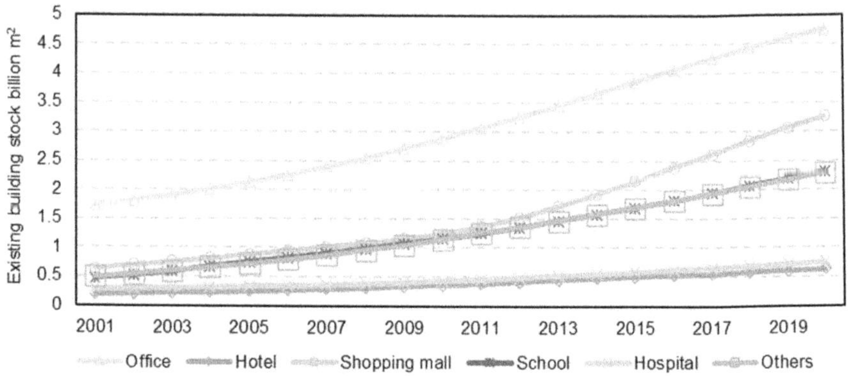

Fig. 3.1 Growth trend of building stock of public and commercial buildings by type

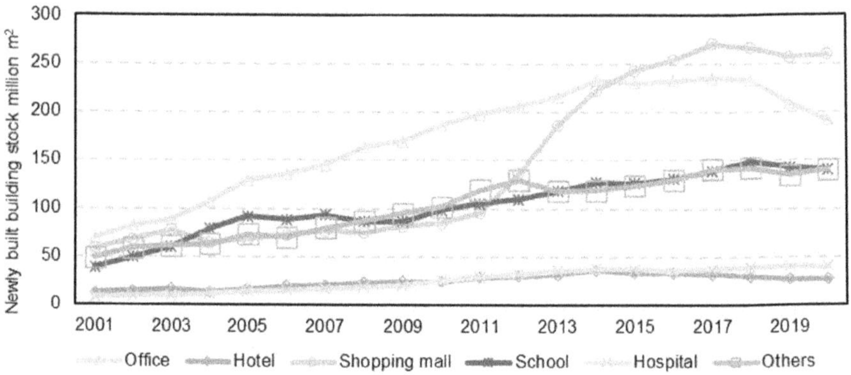

Fig. 3.2 Trend of change in newly built building stock of public and commercial buildings by type

others due to their small stock in early years. However, in recent years, social development has promoted the mushrooming of the buildings stock represented by transportation hub. The number of civil transport airports in China increased from 175 in 2010 to 241 in 2020. The number of urban rail transit stations in operation had doubled from 2,236 in 2015 to 4,681 in 2020. The number of new urban rail transit stations added per year has also shown an increase, rising from 435 in 2016 to nearly 700 in 2020 (Fig. 3.3).

The above trends correspond to the current structure of P&C building stock in China. The results of comparing the per capita P&C building floor area among countries are shown in Fig. 3.4. South Korea, the US, the Netherlands, Denmark, Finland, Germany and Canada have the largest per capita P&C building area, which is more than 20 m^2/capita, followed by Switzerland, New Zealand and Sweden, whose per capita P&C building area is about 15–20 m^2. The per capita P&C building area of Japan, France, the UK and Spain is comparable to that of China and is 5–15 m^2/capita. The per capita P&C building area of India is very small, with a lot of

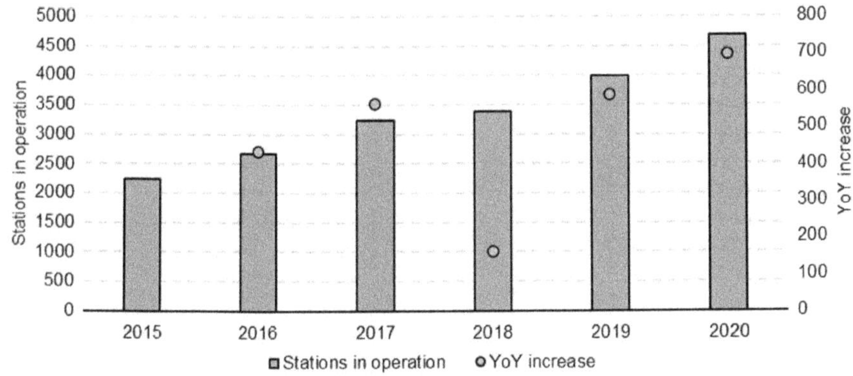

Fig. 3.3 Numbers and annual increases of urban rail transit stations in operation

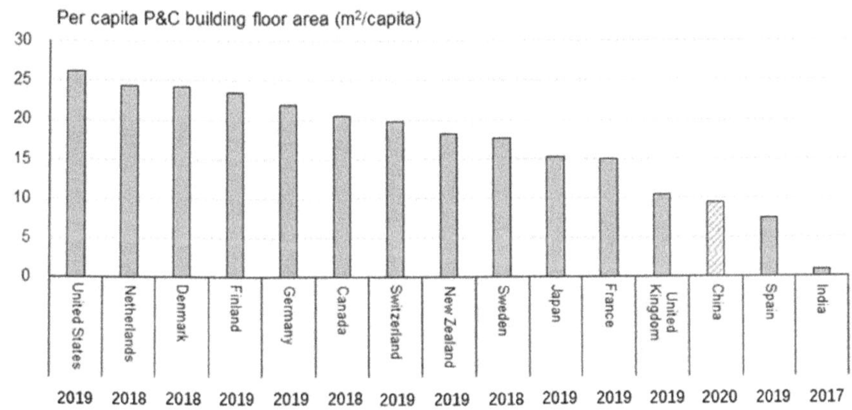

Fig. 3.4 Comparison of per capita P&C building floor area in different countries

room for growth. The results of per capita P&C building area categorized by the type of P&C buildings are shown in Fig. 3.5. At present, the per capita office and mall building floor area in China have almost reached the level of developed countries, with the data in the same order of magnitude as that of France, Spain and the UK. The per capita hotel floor area of China is still much lower than that of developed countries and about one-fifth of that of the US. The per capita school and hospital floor area of China are close to those of some developed countries, but those of the country with the largest per capita area are 4–5 times than those of China, so such types of P&C buildings will become the focus of future development.

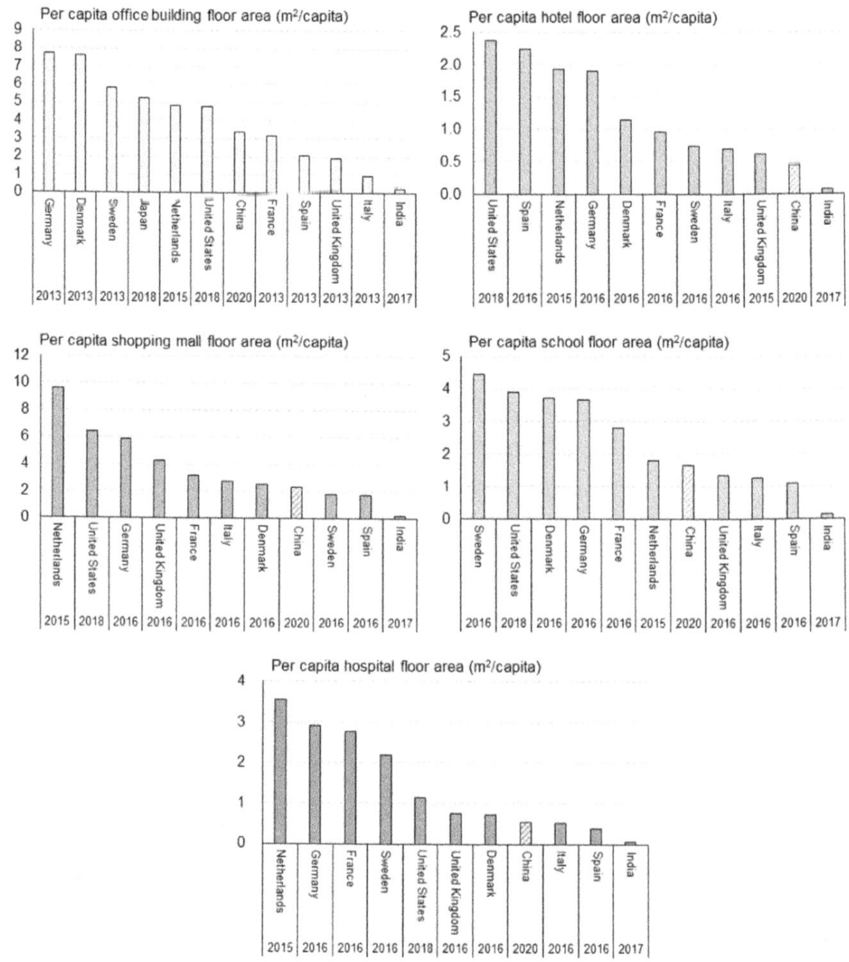

Fig. 3.5 Comparison of per capita floor area of different types of P&C buildings in different countries. *Notes* [1] For the office building floor area data of Japan, the original data indicators are related to "office + shop", so this type includes office and mall buildings. As a result, the statistical standard of Japan is different from that of other countries. [2] The P&C buildings in the US fall into 15 categories, and the data of some types of P&C buildings are merged according to the classification method used in the figure

3.2 Characteristics and Development Trend of the Energy Consumption of Public and Commercial Buildings

3.2.1 Changes in Key Fields of Public and Commercial Building Energy Saving During the 13th Five-Year Plan Period

During the 13th Five-Year Plan period, new changes have been seen in the energy consumption of P&C buildings and the key fields of energy saving. With the implementation of a batch of national science and technology support projects, the technologies, methods and ideas of energy saving in P&C buildings have also changed, with main characteristics as follows:

First, the key fields of P&C building construction and energy saving have changed. As elaborated above, with China's continuous urbanization, the focus of future P&C building development will no longer be office, mall and hotel buildings. Especially due to the impact of COVID-19 and the development of e-commerce and express delivery and logistics, the growth of building area of these three major types of P&C buildings slowed down and their construction scale gradually peaked. Moreover, a certain decrease was observed in the energy use intensity per unit area of these three types of P&C buildings. The first reason for such a decrease is that the actual operational energy consumption was controlled as new P&C buildings put more focus on energy saving. The second one is that the operational energy use intensity fell as more existing P&C buildings had gone through different degrees of energy saving reconstruction during the 13th Five-Year Plan period. For more details, refer to the energy consumption monitoring reports on government office buildings and large-scale P&C buildings as published in Shanghai, Shenzhen, and other cities.

Second, public service facilities closely related to people's lives, such as school/educational buildings and medical buildings as well as transportation buildings including metro stations, high-speed railway stations and airport terminals, developed rapidly during the 13th Five-Year Plan period, and their scale also exhibited very significant growth. However, the energy use intensity per unit area of these three types of P&C buildings showed no significant increase and even decreased gradually during the 13th Five-Year Plan period, partly because of the application of energy saving technologies and partly due to the impact of COVID-19. Besides, the use intensities of such buildings had experienced different levels of decline. It is expected that, by 2035, especially during the 14th Five-Year Plan and the 15th Five-Year Plan periods, the scales of these three types of P&C buildings will grow further and still be the key fields of P&C building energy saving. Energy saving technologies for transportation buildings achieved better development during the 13th Five-Year Plan period. Nevertheless, for school/educational buildings and medical buildings, due to the small scale of each building and scattered investors and construction bodies, both the energy saving reconstruction/renewal of existing buildings and the energy saving of new buildings need to be carried out project by project, and no pattern could be quickly reproduced and popularized yet. As a result, much was said but little was done about energy saving, and the actual progress was slow.

Third, the boundaries between P&C buildings and some "industrial buildings" were gradually blurred, and the field of P&C building energy saving gradually infiltrated into buildings with higher requirements for indoor environment and constructed on "industrial land" or "land for logistics and warehousing". For instance, with the development of e-commerce and logistics and the impact of COVID-19, rapid growth was achieved in low-temperature cold storage, medium temperature cold storage, high-temperature cold storage and normal temperature storage spaces on the land for warehousing and logistics, and the energy saving of such spaces was receiving increasing attention. For such clean workshops or constant temperature and humidity workshops as required for high value-added advanced manufacturing industries, including electronics product manufacturing, food production and biopharmaceutical industry as well as battery manufacturing, there was a large amount of cold water, hot water and low-pressure steam consumption, as well as electricity consumption by fans, water pumps and air compressors, in their environmental control and technological processes. The principles and technical means of energy saving in such processes and environmental controls were very similar to those in P&C buildings. The industrial parks, office spaces and data centers of Internet companies including Tencent, Alibaba, Huawei and Baidu had considerable energy consumption, revealing the huge potential for energy saving. Such "industrial buildings" were more like P&C buildings and were neglected in the past due to the small proportion of their operational energy consumption costs in revenues. However, as the "dual carbon" (carbon peak and carbon neutrality) goals have been deeply rooted among the people and profits have fallen due to the impact of COVID-19, such buildings will become the focus of energy saving during the 14th Five-Year Plan period.

Fourth, the focus should be put on cultural and sports P&C buildings with special environmental requirements because they have great potential for energy saving. For example, the venues of the 2022 Winter Olympics and such sports stadiums and Olympic Sports Centers that have been built across China in recent years in preparation for the Asian Games, the FISU World University Games, the National Games of the People's Republic of China, and provincial sports games. These facilities are of high construction standards and require large cooling and heating power supplies as well as large HVAC and electromechanical systems. The events are not held frequently, so it is very common for equipment and facilities to operate at low load and low efficiency. In particular, for indoor swimming pools, skating centers and even indoor skiing sites, there are corresponding "technological requirements" and the energy consumption is higher. Due to the high environmental requirements of exhibits, many museums adopt a four-pipe air conditioning system, which operates 24 h a day all year round with a serious problem of cold and heat offset. Similar P&C buildings with high requirements for the indoor environment include botanical gardens and aquariums, which also have great potential for energy saving.

Fifth, new opportunities for energy saving and low-carbon development in communities, campuses and new districts of mixed functions and composed of multiple buildings. In the past, a "regional energy system" was often associated with the so-called "energy saving technologies" including combined cooling, heating and power (CCHP) and district cooling. However, through the development and engineering practice over the recent two decades, and the test with operation effects, actual energy costs and other real data, the CCHP system, district cooling system and other extensive technologies which were once suitable for the building use patterns and lifestyles of the US have proven to be unfit for the actual demands of communities, campuses and new districts in China. Since its founding in 2005, the Building Energy Research Center (BERC), Tsinghua University, has continually refuted the applicability of such systems and technologies in China with measured data and theoretical analysis in the annual academic week of building energy saving and the *Annual Development Research Report on Building Energy Efficiency in China*. The practice has proven that they are unfit for China's demands, not because of "economic underdevelopment" or "unadvanced concept" of China, but because the buildings in China, no matter what types they belong to, do not follow the US pattern of "full-time and full-space heating and air conditioning" and "100% mechanical ventilation". In the final analysis, it is because of the culture of Chinese people, buildings and cities formed over thousands of years or because the inherited "DNAs" are "low-carbon". For communities, campuses and new districts of mixed functions and composed of multiple buildings, to achieve the "dual carbon" goals, photovoltaic power generation and all kinds of zero-carbon low-grade heat sources are used to establish a flexible, scattered, coordinated and shared energy utilization and conversion system. Besides, the Internet, the Internet of Things (IoT) and other new technologies are used for smart operation, maintenance, regulation and control, thus it is possible to achieve "zero-carbon operation" for the communities, campuses and new districts. During the 14th Five-Year Plan period and the 15th Five-Year Plan period, a batch of demonstration projects of zero-carbon campuses will be tested by practice, and then popularized and reproduced.

Furthermore, the most important change is the "30/60 dual carbon goals" (achieving carbon peak in 2030 and carbon neutrality in 2060) proposed by President Xi Jinping at the 75th session of the United Nations General Assembly on September 22, 2020, which is a major strategic deployment for China to strengthen the development of ecological civilization and implement the new development philosophy in a comprehensive, complete and accurate manner. P&C buildings are the key to building energy saving. Besides the control over total energy consumption and energy use intensity, the total carbon emission and carbon intensity should also be controlled. The focus should be shifted from furthering energy saving and improving building and system energy efficiency to improving the flexibility and tenacity of building energy use systems. PEDF (photovoltaic, energy storage, direct current and flexibility) buildings and community and campus energy systems should be constructed to improve renewable energy utilization and absorb the renewable electricity from power grids. Apart from establishing rules, regulations and standard systems, more scientific and strict measures should be implemented. In addition to the requirements of "energy saving design", a whole-process energy saving and low-carbon management system should be established to cover project initiation, planning, design, procurement through tendering and bidding, installation and construction, commissioning and acceptance, as well as continuous operation, maintenance and update, so as to significantly reduce the carbon intensity of P&C buildings as soon as possible. These are intended to "decouple" total carbon emissions of P&C buildings from the growth of building area; continuously reduce the actual operational energy consumption and carbon emissions of P&C buildings, as well as carbon emissions in energy utilization in new P&C buildings and communities, campuses and new districts during the 14th and the 15th Five-Year Plan periods; realize the goals of "early carbon peaking, short plateau phase, tail-up tendency prevention, true innovation and steady decrease", and assist the urban and rural construction sector in achieving the "dual carbon" goals.

3.2.2 *Characteristics of Changes in Energy Use Intensity of Main Types of P&C Buildings*

Since 2007, many provinces and cities in China have established energy consumption monitoring platforms for government office buildings and large-scale P&C buildings. During the 13th Five-Year Plan period, investigation and survey were conducted on the P&C building energy consumption monitoring platforms of Shanghai, Shenzhen, Qingdao and Xiamen and of Changning, Huangpu and Xuhui Districts in Shanghai, as well as on the energy management system monitoring platforms of three corporations including China Resources Land, Grandjoy Holdings and SOHO Group. In particular, the platform data application effects were evaluated, the difficulties and problems encountered in the analysis and application of data from the energy consumption monitoring platforms were sorted out, besides, on this basis, the *Information Presentation Standard for Energy Use Data and Influencing Factors of Public Buildings and Systems* was presented, hoping to improve the availability, activity and customer stickiness of energy consumption monitoring data of P&C buildings. At present, the abovementioned platforms are still in normal operation and subject to continual improvement. However, due to the impact of COVID-19, few data and functional updates of these energy consumption monitoring platforms have been made since 2020. The following is an overview of the characteristics of changes in energy use intensity of main types of P&C buildings from the latest reports on the status of the energy use monitoring platforms for P&C buildings published by Shanghai and Shenzhen. The main types of P&C buildings here include government office buildings, commercial office buildings, mall buildings, hotel buildings, cultural and educational buildings, medical buildings, sports buildings, and comprehensive buildings.

(1) Shanghai

On August 23, 2021, the *2020 Report on Energy Consumption Monitoring and Analysis for Government office buildings and Large-scale Public Buildings of Shanghai* was officially released. The report was prepared by the Shanghai Municipal Commission of Housing and Urban–rural Development and the Shanghai Municipal Development and Reform Commission. Based on the data of the Energy Consumption Monitoring Platform for Government office buildings and Large-scale Public Buildings of Shanghai (referred to as the "Energy Consumption Monitoring Platform"), the report analyzed the energy consumption monitoring situations and characteristics of operational energy use for government office buildings and large-scale public buildings in Shanghai in 2020 and during the 13th Five-Year Plan period.

Shanghai started to build its Energy Consumption Monitoring Platform in 2010. Through gradual expansion and improvement, it has completed an integral energy consumption monitoring system for P&C buildings including 1 municipal platform, 17 district-level sub-platforms and 1 municipal organ sub-platform. As of December 31, 2020, a total of 2,017 P&C buildings in Shanghai had completed the installation of energy consumption metering devices and achieved data networking with the Energy Consumption Monitoring Platform, with a total building area of 92.083 million square meters. Among them, there were 200 government office buildings, which accounted for 9.9% of the total number of monitored buildings and had a total building area of 3.692 million square meters; and there were 1,817 large-scale P&C buildings, accounting for 90.1% of the total number of monitored buildings and with a total building area of 88.391 million square meters. During the 13th Five-Year Plan period, a total of 729 buildings were newly connected to the Energy Consumption Monitoring Platform, with an additional building area of nearly 35 million square meters. Based on this platform, and with support from the National Science and Technology Support Program, Shanghai Research Institute of Building Sciences Co., Ltd. had developed and upgraded it into the national big data platform for energy consumption of green buildings.

According to relevant statistics, in 2020, the annual total electricity consumption of P&C buildings connected to the Energy Consumption Monitoring Platform in Shanghai was about 8.55 billion kWh, equivalent to about 6.737 million tons of CO_2 emissions. According to statistics for different types of buildings, the total electricity consumption of office buildings, mall buildings, comprehensive buildings, tourist hotel buildings and other traditional P&C buildings was greater, with a higher proportion, see Fig. 3.6. However, due to COVID-19, the annual electricity intensity, i.e., annual average electricity consumption per unit area, was 93 kWh/m^2 in 2020. Due to such factors as shortened operation duration and reduced flow of people in buildings during the pandemic, the energy consumption of public buildings from January to April 2020 was less than that in the same period in 2019. Nevertheless, with the resumption of work and production and the effective control of the pandemic, the energy consumption showed an obvious increase from May 2020 and bounced back to a level similar to that in the same period in 2019.

The Fig. 3.7 below shows the electricity intensities of P&C buildings connected to the Energy Consumption Monitoring Platform in 2020. The electricity consumption data of cultural buildings, sports buildings and other buildings are for information only due to the limited number of uploaded data samples. It can be seen that mall and medical buildings still have the highest energy use intensities, while educational

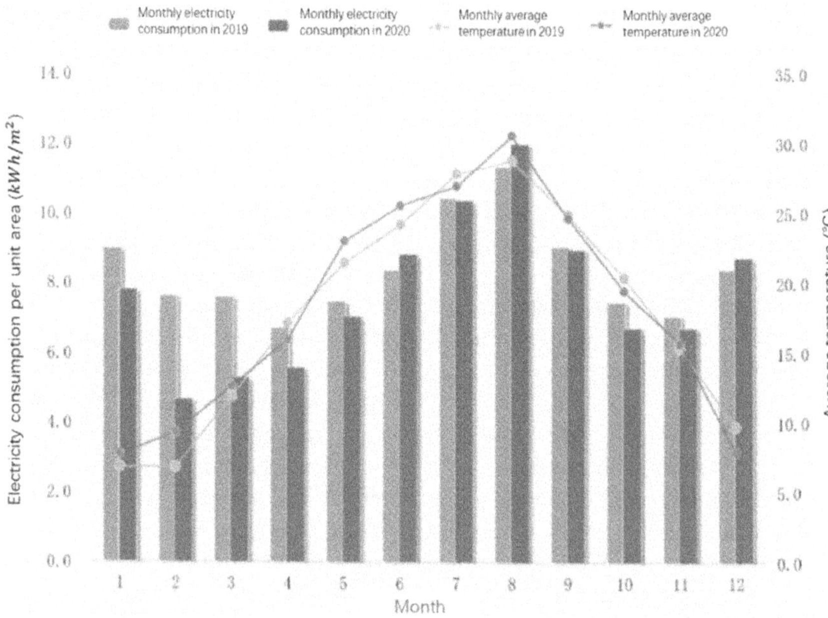

Fig. 3.6 Monthly electricity consumption of connected commercial buildings in 2019–2020

buildings have the lowest energy use intensity. Among cultural and sports buildings, the actual energy use intensity varies significantly with different subtypes of P&C buildings, and further statistical analysis of subtypes may be performed in the future.

In addition, in terms of changes in the overall energy use intensity and the total energy consumption of P&C buildings during the 13th Five-Year Plan period (Fig. 3.8), the total energy consumption of P&C buildings was close to the peak, the energy use intensity fluctuated without any significant increase, and the energy use intensities of some types of P&C buildings somewhat decreased, indicating that the energy consumption per unit area of P&C buildings was under effective control during the 13th Five-Year Plan period (Fig. 3.9). This owes much to the continual progress of energy consumption monitoring, energy audit, energy-saving retrofit, energy efficiency improvement and other supervisory works for P&C buildings in Shanghai, as well as to the raising awareness of energy saving among building owners.

It can be seen from the works above that, although energy saving of P&C buildings can be only carried out project by project, perseverance can yield fruits. During the 14th Five-Year Plan period, with the implementation of carbon peaking action plans for the state and the cities, it is expected that the energy use intensities and carbon intensities of P&C buildings will decrease significantly. Let's wait and see.

(Note: The above charts are taken from the *2020 Report on Energy Consumption Monitoring and Analysis for Government office buildings and Large-scale Public Buildings of Shanghai*).

(2) Shenzhen

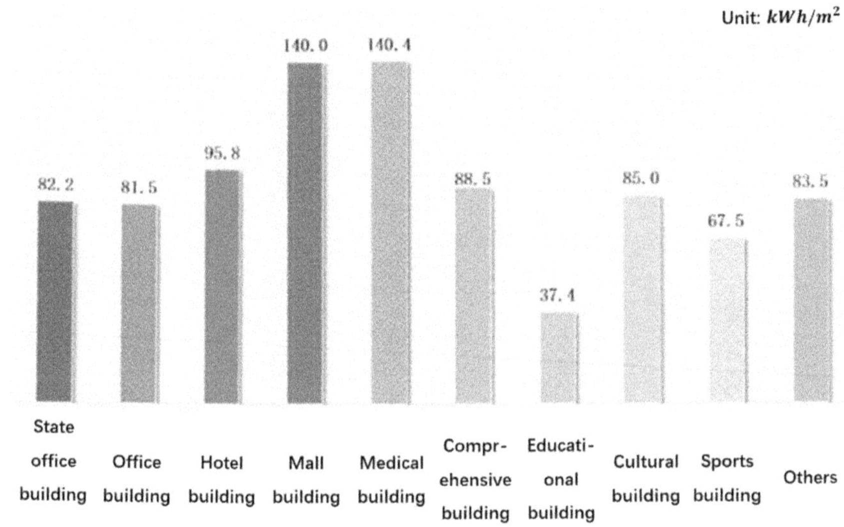

Fig. 3.7 Monthly electricity consumption of different types of connected commercial buildings in 2019–2020

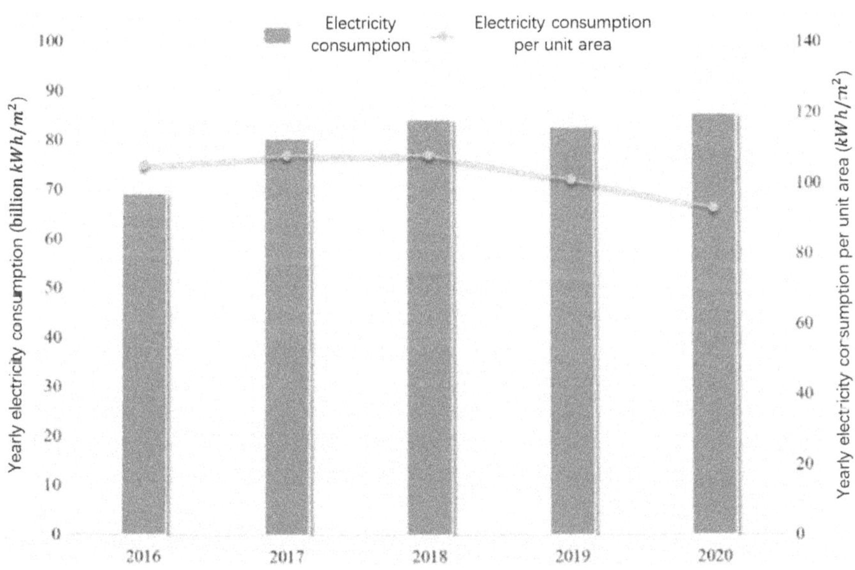

Fig. 3.8 Changes in electricity consumption of connected public buildings over the years during the 13th Five-Year Plan period

Calculations were made according to the *Shenzhen Statistical Yearbook 2020* and the data released by the Shenzhen Power Supply Bureau of China Southern Power Grid. In 2020, the total electricity consumption of P&C buildings in Shenzhen was

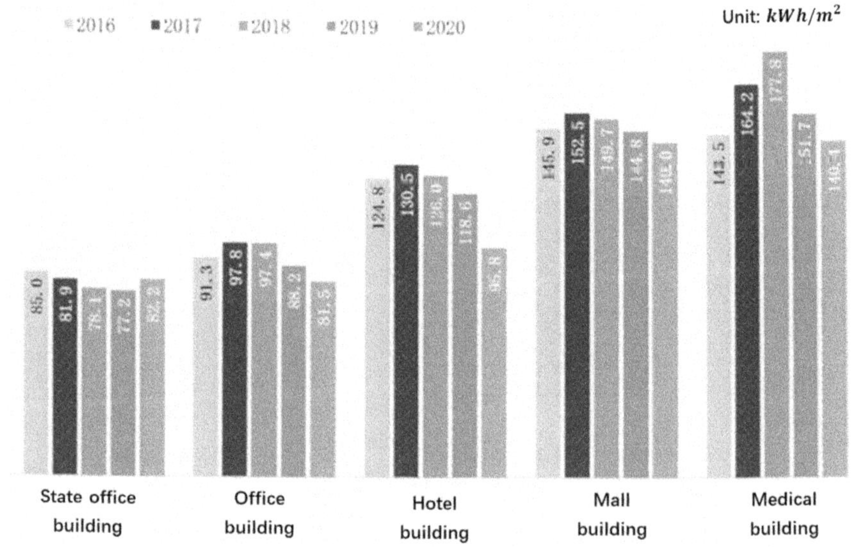

Fig. 3.9 Changes in electricity consumption of different types of connected public buildings over the years during the 13th Five-Year Plan period

approximately 27.37 billion kWh, accounting for about 28% of the total electricity consumption by the whole city. Energy saving of P&C buildings is particularly important under the new situation of carbon peaking and neutrality. As of January 2021, a total of 702 government office buildings and large-scale P&C buildings had been connected to the Energy Consumption Monitoring Platform of Shenzhen. After the deduction of the buildings for which the monitoring had been canceled or had lasted for less than a year due to urban renewal, building demolition, or other reasons, the total number of buildings monitored in 2020 was 659, with a total building area of 34.56 million square meters. Compared to 2019, the number of monitored buildings increased by 83 and the area of such buildings increased by 7.73 million square meters.

In terms of energy use intensity, the electricity intensity per unit area of P&C buildings monitored in Shenzhen in 2020 was 96.5 kWh/m^2. The table below shows the energy use intensities of the main types of P&C buildings monitored in Shenzhen in 2020, the unit is kWh/m^2·a.

The energy use intensities of the main types of P&C buildings in 2019 are compared with those in 2020 in Fig. 3.10. It can be seen that the energy use intensities of P&C buildings have a year-on-year decrease by about 11.5% from 109.0 kWh/m^2 in 2019, and P&C buildings with higher energy use intensities including malls and hotels were significantly affected by COVID-19 in 2020. In the report, the monthly electricity consumption and monthly outdoor mean temperatures in 2019 were compared in detail with those in 2020, as shown in Fig. 3.11.

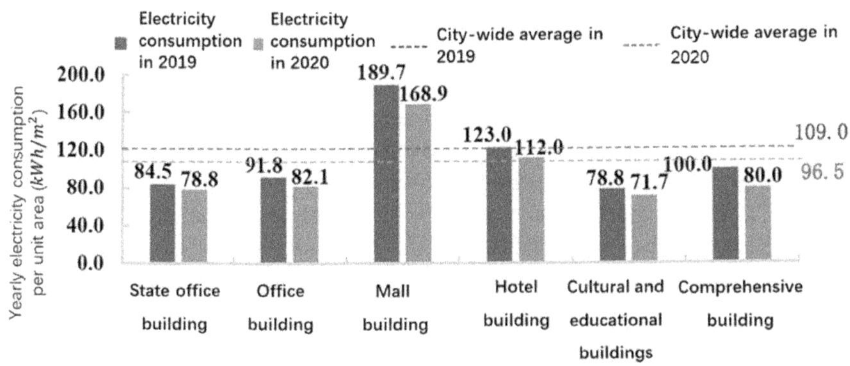

Fig. 3.10 Electricity consumption of monitored building in Shenzhen

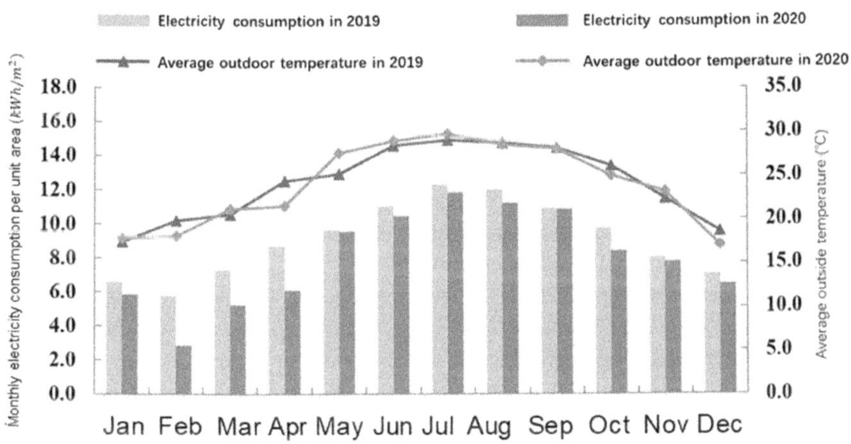

Fig. 3.11 Comparison of electricity consumption of monitored public building in Shenzhen from 2019 to 2020

The characteristics of energy consumption of all types of P&C buildings in Shenzhen can also be observed from the *Report on Energy Consumption Monitoring of Large-scale Public Buildings in Shenzhen (2020)*. For example, the following figure gives the energy consumption analysis for P&C buildings in ventilation, air conditioning and dehumidification seasons. The ventilation season lasts 181 days from October 8 to the next April 5, while the air conditioning season lasts 135 days from May 26 to October 7. The dehumidification season lasts 50 days from April 6 to May 25.

It is observed that the energy use intensities of different types of P&C buildings are particularly relevant to ventilation and air conditioning modes. Government office buildings have the lowest electricity consumption in the ventilation season, as a result of the much natural ventilation. However, the commercial office, hotel and mall buildings have little difference in electricity consumption in the air conditioning and ventilation seasons, indicating that the utilization rate of natural ventilation should be improved. The electricity consumption per square meter per day in each season can be obtained by dividing the electricity consumption of each of the three seasons by the number of days. For instance, from the perspective of the average of all types of P&C buildings (Fig. 3.12), the average daily electricity intensity in ventilation season is about 220Wh/m^2·d, which reflects the "basic electricity consumption" except for electricity consumption by refrigeration and air conditioning in P&C buildings, i.e., the fairly stable electricity consumption by lighting and electrical equipment as well as elevators. The average daily electricity consumption in the air conditioning season is 330 Wh/m^2·d, and the excess portion, i.e., 110 Wh/m^2·d, is associated with the electricity consumption of the refrigeration and air-conditioning system. If the electricity consumption of the refrigeration and air-conditioning system, including cold source, transmission and distribution system and terminal fan, is estimated by a system efficiency of 3.0, the average daily cool loss of buildings is about 330 Wh_{cool}/m^2·d. Assume the air conditioning refrigeration is turned on for about 8 h, then the calculated average cooling load is about 40 W/m^2, which reflects the situations of government office buildings, commercial office buildings and comprehensive buildings in the figure below. For mall buildings, the average daily electricity consumption in air conditioning season is about 1.8 times the average value of all types of P&C buildings, i.e., approximately 640 Wh/m^2·d, which indicates that the electricity consumption of the refrigeration and air-conditioning system is about 420 Wh/m^2·d and the average daily cool loss is about 1,260 Wh_{cool}/m^2·d. Considering that air conditioners in malls are usually turned on for 12–14 h, the average cooling load (or cooling capacity) is about 100 W/m^2. The cooling capacity of hotel buildings can also be analyzed similarly, but it should be noted that the air conditioning systems of hotel buildings generally operate continuously for 24 h; therefore, the average daily cooling load per square meter is only about 35 W/m^2.

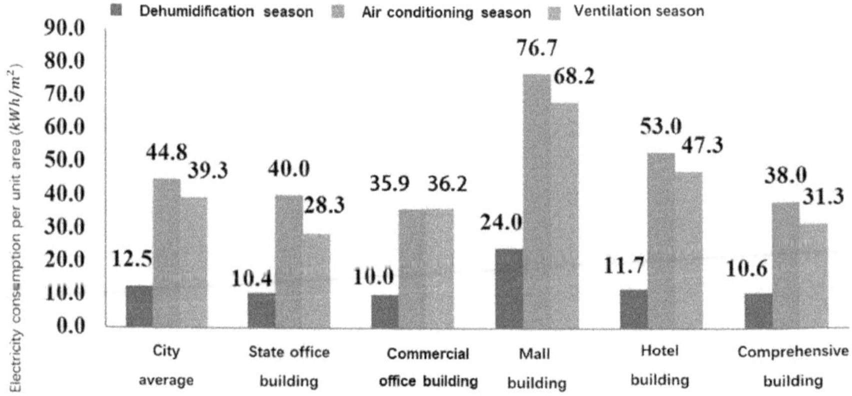

Fig. 3.12 Electricity consumption intensity of monitored public buildings

It is illustrated that, based on the year-by-year research and analysis for the energy consumption monitoring data of P&C buildings of the whole city, Shenzhen has further expanded the breadth of data and constantly leveraged the value of the data to attract more P&C buildings to connect to the Energy Consumption Monitoring Platform. On the one hand, Shenzhen Municipal Housing and Construction Bureau has issued a series of policies and standards concerning energy consumption monitoring and building energy saving to standardize the energy consumption data access. On the other hand, to make the platform more convenient to use and improve the building energy saving management of government and businesses and the public's energy consumption management experience, an applet for the platform has been rolled out to enable real-time viewing and management of the building energy consumption in the hand. The applet has three different clients, i.e., administration client, property management client and user client, for free access by the municipal and district-level government administrators, building owners/property management, and building tenants/users. This development strategy based on market mechanism and customer stickiness is worth promoting and replicating in other cities.

3.2.3 Changes in the "Bimodal Distribution" Characteristics of Energy Use Intensity of P&C Buildings

Around 2010, through preliminary analysis of the energy consumption statistics and monitoring platform data of P&C buildings in China since 2007 and based on the comparison with the statistical results of energy consumption, especially the statistical results of energy use intensity, of P&C buildings in the US, Japan, European countries and other developed countries, Xiao He et al. pointed out that, as to the energy use intensity of P&C buildings in China, both the energy use intensity per unit area and the per capita energy use intensity were far below those of the developed countries. This phenomenon is caused by the historical stage and characteristics of China's social and economic development, and the energy use intensity of P&C buildings exhibits an obvious "bivariate distribution structure" or a "bimodal distribution", as shown in the figure below.

Taking office buildings for example, the energy use intensity of most office buildings is concentrated on a lower level of 50–70 kWh/m^2·a, while that of the other small part of buildings at a higher level of 120–150 kWh/m^2·a, which is 1.8–2.6 times the former. The above viewpoint was first proposed by Jiang Yi, Academician of the Chinese Academy of Engineering, in 2003, namely, P&C buildings in China should be classified by area into two types for separate treatment, and large-scale P&C buildings with higher energy use intensity shall be the focus of energy saving (Fig. 3.13). In recent years, more energy consumption data has revealed the difference in the distribution of energy consumption of the two types of P&C buildings, and continually explained the reasons for the bivariate distribution of energy use intensity. One reason is that the use intensity varies with P&C buildings. The more important reason is that

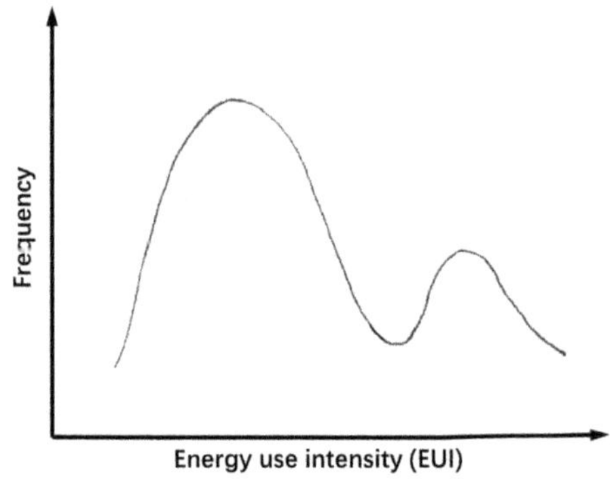

Fig. 3.13 Energy use intensity distribution

the indoor environment construction concepts and methods are different, leading to the formation of two types of P&C buildings with different energy consumptions. This viewpoint was adopted in the *Standard for Energy Consumption of Civil Buildings* issued and implemented in 2016, which gives the constraint values and guide values of energy use intensities for Type A and Type B P&C buildings respectively, corresponding to the P&C building types represented by the two peak frequencies in the above bivariate distribution.

In recent years, especially during the 13th Five-Year Plan period, more statistical results from energy consumption survey of P&C buildings have shown that, at present, the energy consumption of P&C buildings in China still has a bivariate distribution, in which the energy use intensity of Type B large-scale P&C buildings on the right shows no significant increase and that of Type A general P&C buildings on the left shows no significant decrease, and "the average value or median of energy use intensity of all P&C buildings" is still not typically representative. However, urban construction, in particular urban renewal, has led to the demolition of a batch of Type A general P&C buildings for reconstruction and expansion into Type B large-scale P&C buildings. Hence, there arises the pressure of growth in the energy use intensity of P&C buildings as a whole. This necessitates strengthening energy saving supervision, raising energy saving standards and strictly taking actual operational energy consumption and carbon emissions as the indicators for appraisal and examination, so as to promote the continuous reduction of energy use intensity and carbon intensity of new P&C buildings and the reduction of energy use intensity per unit area and carbon intensity per unit area in the actual operational process through continuous energy saving and low-carbon renewal and reconstruction as well as energy efficiency improvement of existing P&C buildings.

The ongoing research and studies over the past two decades show that the construction of P&C buildings and their indoor environment in different cultures, lifestyles and concepts will lead to a huge difference in energy consumption. In essence, the total energy consumption of P&C buildings in a country is determined by the proportions of the two types. With the social and economic development, especially with the support of such concepts as common prosperity and sharing of development dividends, the social economy and the energy use intensity of P&C buildings generally transform from "bivariate distribution" to "univariate distribution", because the difference between two peaks mirrors the difference in the system, whereas the one-peak distribution reflects the difference in technical and management factors from within, which should be an inevitable trend of historical development.

It is estimated that, in the next decade, the growth in the scale of Type B large-scale P&C buildings on the right will slow down, the energy use intensity per unit area will gradually decrease, the number of P&C buildings with higher energy use intensities will gradually decline, and the peak on the right will move leftwards. In the meantime, a batch of general P&C buildings with lower plot ratios and lower use intensities can also gradually reduce energy consumption and carbon emissions, i.e., the left "peak" keeps moving leftwards. In the future, we should not only track the changes in the distribution law of energy use intensity of P&C buildings, but also increase the tracking and analysis for the distribution law of carbon intensity of P&C buildings. We should timely identify and understand the characteristics of changes in the energy use intensity, carbon intensity, use intensity and scale of P&C buildings in China, in different cities and of different types, so as to make sure P&C buildings can meet the people's aspiration to live a better life and also achieve the carbon peaking and neutrality goals on schedule. This will be the main task at the next stage of P&C building energy saving.

Chapter 4
Decarbonize Public and Commercial Buildings

4.1 Role of Public Buildings in Achievement of the "Dual Carbon" Goals

(1) **Impact of energy structure reform on energy demand**

The energy structure reform is a fundamental link that must be implemented to achieve the carbon neutrality goal. One of the main tasks of implementing the carbon neutrality strategy is to transform a carbon-based power system that relies on fossil energy into a zero-carbon power system that is based on renewable energy sources. Table 4.1 gives the power supply composition of current power systems in China and that of the zero-carbon power system which is expected to be realized in the future. It shows that the proportion of the installed capacity of wind and solar PV power will increase from 20% currently to 80% and that the proportion of electricity provided by wind and solar PV power will increase from less than 10% currently to approximately 60%.

In addition to the issue of installation space of wind and solar PV power, the asynchronous power change of wind and solar PV power with that of terminal electricity utilization is the main bottleneck encountered in the large-scale development of wind and solar PV power. Figure 4.1 shows the changes over time in wind and solar PV power within one week, which are mainly characterized by volatility and uncertainty. "Source changing with load" is the basic regulation mode for the traditional electric power system dominated by fossil energy and hydropower. Any change on the load side needs to be balanced by real-time regulation on the power source side, while any change in the regulation process needs to be balanced by the huge moment of inertia of the generator rotor system. After wind and solar PV power becomes a main power source, their generated outputs will depend on weather conditions at the time and will be very difficult to regulate unless wind and solar PV power is curtailed, thus leading to stricter requirements on the energy use system and users in this mode of energy supply. The natural volatility and change of wind and solar PV power determine that changes on their supply sides are much greater than those in the supply mode of the

Y. Jiang et al., *Decarbonize Public and Commercial Buildings*,
https://doi.org/10.1007/978-981-19-5526-6_4

Table 4.1 Installed capacity and generating capacity of China's electric power systems in 2019 and 2050

	Present situation (2019)		Planned status 2050	
	Installed capacity (100 million kW)	Annual total generating capacity (trillion kWh)	Installed capacity (100 million kW)	Annual total generating capacity (trillion kWh)
Hydroelectric power	3.8	1.6	5	2
Nuclear power	0.5	0.4	2	1.5
Wind and solar PV power	4	0.55	60	8
Peak shaving thermal power	11	5	6.5	1.5
Total	19	7.5	74	13

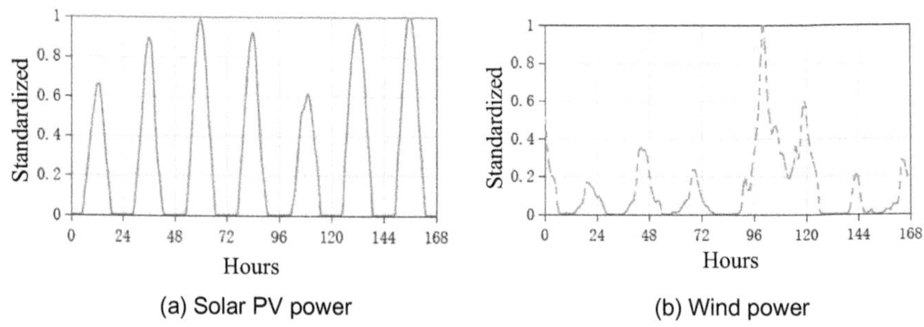

Fig. 4.1 Changes over time in generated outputs of wind and solar PV power (with a typical week of July in Beijing as an example)

traditional energy system. Hence, it is necessary to develop a flexible response and system regulation mode in which "load changing with source" and demand is guided by supply, or to introduce a huge energy storage link to balance the power difference between source and load. The reform of energy structure necessitates the establishment of a system adaptable to the new relationship between supply and demand in the future energy structure.

Wind and solar PV power receives the wind and solar energy from nature and their generated outputs are almost in direct proportion to floor areas. According to the current technology level of wind and PV power generation, the generating capacity per unit horizontal area is about 100 W/m^2, far lower than that of nuclear, thermal and hydroelectric power. In the light of the plan in Table 2.1, if China needs to install 6 billion kW of wind and solar PV power in the future, about 60 billion square meters of horizontal space will be required for the installation, which is equivalent to approximately 100 million mu (1 mu = 0.0667 hectare) of land. China needs 1.8

billion mu of farmland to ensure the basic food supplies. By contrast, the additional 100 million mu of land for energy purposes are huge.

Thus, it is natural to think of developing wind and solar PV power in a large number of sandy and Gobi deserts in western China. This indeed is an important trend in the development of wind and solar PV power in recent years. However, the main electrical load in China is concentrated in the east of the Hu Huanyong Line, while the sandy and Gobi deserts available for large-scale development are located west of the Hu Huanyong Line, with a distance of over one thousand kilometers between the east and the west. For this reason, China has built multiple superpower long-distance transmission lines in recent years. However, to make these lines function effectively and ensure the stability of power transmission, it is necessary to "bundle" hydropower or thermal power together with wind and solar PV power to supply fairly stable power. Based on the change rule of wind and solar PV power in a day, the ratio of required hydropower or thermal power to transmitted wind and solar PV power shall reach 1:1 at least. Although there are abundant water power resources in western China, its total quantity is no more than 500 million kW. Therefore, "bundling" can be done for 500 million kW of wind and solar PV power only. More wind and solar PV power needs to be matched by local coal- and gas-fired thermal power. In this way, it is impossible to lower the proportion of coal- and gas-fired thermal power in future electric power systems, and thus impossible to achieve the goal of zero-carbon electricity. Certainly, huge energy storage facilities can be set up in wind and solar PV power bases to stabilize the wind and solar PV power of a day through adjustment by energy storage. In this case, the required energy storage capacity shall reach 50–60% of the daily total generating capacity at least. The generating capacity of 1 W of solar PV is 10 Wh per day, so the required power storage capacity will be 5–6 Wh. Now, a chemical energy storage device costs more than RMB 6, while the cost of 1 W of PV device is less than RMB 1.5 and that of a PV power generation system, including support brackets, inverters and transformers, is no more than RMB 4/W. Thus, the chemical energy storage adopted in the place of origin for electricity will increase the system cost from RMB 4/W to RMB 10/W. On the other hand, solar PV electricity is characterized by high power in the daytime and zero power at night, which approximates the load characteristic of terminal electricity consumption in eastern China. If the power transmission from west to east is constant in a day, the eastern region needs to accumulate and store the extra electricity at night for consumption in peak hours in the daytime. According to the change rule of typical electricity consumption by office buildings in a day, the quantity of electricity to be accumulated and stored at night is about 30–40% of the total electricity consumption of the very day. This means that huge energy storage resources are required to regulate the difference between peak and valley on the consumption side and that the chemical energy storage cost in the eastern region will increase by RMB 3–4 for 1 W of PV electricity transmitted from the western region. After storing energy twice, the cost of chemical energy storage is almost 3 times that of the PV power generation system.

If solar PV power is developed in load-intensive areas of eastern China, there will be a certain degree of overlapping between the change of solar PV power and that of electrical loads of buildings in a day. In this case, the chemical energy storage capacity corresponding to 1 W of PV is only 2–3 Wh, which is much lower than the total energy storage capacity (8–10 Wh, including 5–6 Wh in the west and 3–4 Wh in the east) required for installation in the western region. In this way, the energy storage capacity required for the PV system can be reduced significantly, and the quantity of electricity stored and discharged is only 1/3 of that in the power generation mode in the west. The decrease of loss in electricity storage and discharge and in long-distance transmission can improve the system efficiency by 20–30%, which almost offsets the advantage of the west where the solar radiation intensity is 30–40% higher than that in the east. Therefore, it may be more suitable to prioritize the development of solar PV in the eastern load-intensive areas.

As there is an acute shortage of land resources in eastern China, where should PV be installed? Perhaps in all types of empty space, such as rooftops of buildings. According to conclusions drawn from statistical analysis of high-resolution satellite images and of on-site sampling surveys, the converted available horizontal surface areas of rooftops in urban and rural areas of China are about 41.2 billion square meters. After full consideration of actual difficulties in installation and allowing for unforeseen circumstances, it is concluded that 830 million kW of PV can be installed on urban empty rooftops across China, with an annual generating capacity of 1.23 trillion kWh; and 1.97 billion kW of PV can be installed on rural empty rooftops, with an annual generating capacity of 2.95 trillion kWh. Thus, urban and rural areas can install a total of 2.8 billion kW of PV, which is more than 70% of China's planned total installed capacity of PV in the future, and the potential generating capacity is 4.2 trillion kWh, over 70% of China's planned total generating capacity of PV in the future. Hence, PV power generation on urban and rural building rooftops and other surfaces exposed to solar radiation should be the main direction for China's large-scale development of PV power generation in the future.

Figure 4.2 shows the equilibrium diagram of future electric power systems in China, including eastern and western regions. The future electricity demand will be 13.7 trillion kWh, including the demands of the western region, the rural areas in the mid-eastern region, the industry, municipal administration and transportation (excluding private and official passenger vehicles) in the mid-eastern region, and the buildings and automobiles (only including private and official passenger vehicles) in the mid-eastern region. The power generation side includes wind power, solar PV power, hydropower, nuclear power, as well as biomass and coal- and gas-fired peak shaving thermal power. The lower part of the diagram indicates the working status of conventional electric power systems. As shown by the dotted boxes, hydropower is used for peak shaving of concentrated wind and solar PV power, and then it is packed to form a manageable power source entering a large power grid. The electricity demands of various loads are met with the common support from their own power supplies and the power supplies from adjacent wind and solar PV power and large power grids. Among them, the demands of the western region can be met by the wind/solar PV power of buildings as well as the concentrated wind/solar PV power and hydropower, which can transmit 0.9 trillion kWh of electricity to the eastern region. In the mid-eastern region, the electricity demands of rural areas can also be met by rooftop PV, which can transmit 0.8 trillion kWh of electricity to

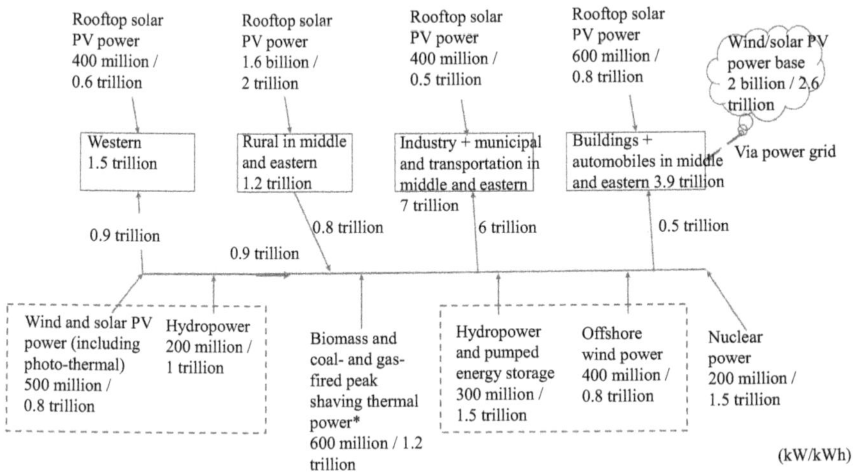

Note*: The emissions from biomass, coal-fired and gas-fired peak shaving thermal power recovered by CCUS are 800 million tons of CO2, with the remaining emissions smaller than those from biomass fuels

Fig. 4.2 Equilibrium diagram of future electric power systems

power grids. The electricity for industry, municipal administration and transportation (excluding private and official passenger vehicles) is mainly guaranteed by power grids. Most demands of buildings and passenger vehicles can be met by the wind and solar PV power surrounding the buildings, and a small part of their demands is met by power grids. To sum up, the demands of the western region and industry are met by hydropower and nuclear power in power grids, while those of the rural areas, buildings and passenger vehicles in the mid-eastern region are mainly met by wind/solar PV power. Therefore, for buildings and passenger vehicles, the imbalanced power changes between supply and demand need to be solved by themselves. This also shows that China can achieve zero-carbon electric power systems with a high proportion of wind and solar PV power in the future, and thus solve the key problems in the energy revolution and carbon neutrality strategy as long as the electricity for urban and rural buildings and for charging passenger vehicles is provided by rooftop PV and the mismatch in time of power changes between supply and demand is addressed. Hence, PV development and flexible electricity consumption in urban and rural buildings are not only important tasks to be fulfilled in the low-carbon development of buildings, but are also of strategic significance in the low-carbon transition of China's entire energy system and in the achievement of carbon neutrality.

(2) **Important roles of public buildings in future energy structure system**

In terms of energy use intensity, the energy consumption per unit area of public buildings is the highest and obviously higher than that of any other civil buildings including urban and rural residential buildings. Similarly, public buildings have the highest carbon intensity among civil buildings. This is closely related to the long life, high energy use intensity, high supportive requirements, multiple functions and complex equipment system of public buildings, and also reflects the important role public buildings play in realizing key objectives, including reducing carbon emissions from buildings and promoting energy saving of buildings.

As to paths of reducing carbon emissions from building operation, first, passive technologies shall be adopted in building design and construction to minimize the demands of buildings for cooling, heating and artificial daylight provided by electromechanical systems. Passive technologies for the energy supply system shall then be adopted to maximize the system's energy supply efficiency. That is to say, to achieve the low-carbon development goal of public buildings, relevant efforts should be made in building design and the building itself, and solutions should also be sought from the building energy use system and the active electromechanical system.

- **For the building itself, passive design** of buildings should be implemented **to reduce demands**. For example, the cooling and heating demands of buildings can be reduced effectively through improvement of envelope performance, which is an important technical means to realize energy saving of building body. Energy saving at the building body and envelope level has earned widespread respect. At present, multiple new technologies of new envelopes and envelope insulation materials have been researched and applied, and targeted technical solutions have been provided to thermal disturbance including the envelope heat transfer and solar radiation heat that affect the building body. The new technology of radiation refrigeration introduced in Chap. 5 of this report is to reduce the unwanted effects of solar radiation by applying new materials, thus enabling the building body to better adapt to the energy saving and low-carbon development requirements. Installing PV on external surfaces of buildings to make full use of available renewables including PV of buildings is also an important technological path that will be widely applied in buildings.
- **The electromechanical system of each building** should follow the principle of "**electrification, distributedness, high efficiency and flexible electricity consumption**". To achieve the goal of building a low-carbon energy system, the building energy use system/electromechanical system should reduce and even avoid the consumption of fossil energy in buildings, and the energy consumption of the building itself should be fully electrified/re-electrified. In the design and operation of the electromechanical system for a building, the building's functional requirements shall be taken into full account. To avoid the waste and unwanted losses resulting from a centralized system, a distributed and adjustable electromechanical system that fits relevant requirements shall be established whenever possible. After the building's functional requirements are met, the role of buildings in the entire energy system shall be repositioned and reframed to better serve the goal of carbon emission reduction. Based on the existing building electromechanical systems that emphasize high efficiency and the equipment systems that seek higher energy use efficiency, buildings shall play their important roles in effectively responding to dispatch requirements of the energy system and in promoting supply and demand matching of the energy system.

Therefore, the low-carbon goals of public buildings shall be achieved through the building itself and the electromechanical system. The building design or the building body shall focus on "**reducing demands and tapping more sources**", namely, the energy use demands of buildings shall be minimized and the use of available renewables including PV of buildings shall be maximized. In the design and operation of the building energy use system or electromechanical system, the principle of "**electrification, distributedness, high efficiency and flexible electricity consumption**" shall be followed as the guide for the design and improvement of the electromechanical system of a public building to better serve the low-carbon goal of the entire urban energy system.

"**Passive first to reduce demands and active in optimization to improve efficiency**" is an important foundation for reducing building energy use demands and carbon emissions from building operation, and this is also the key to energy saving of buildings. On this basis, the full electrification or re-electrification of buildings will serve as an important measure for emissions reduction in the building sector.

Public buildings, as an important link of full electrification in the building sector, are convenient and easy to be fully electrified. For example, the heating demands of public buildings can be met by various heat pumps or centralized heat sources. The heat pump has become an efficient heat supply equipment as a result of lowered requirements for heat source grade with the support of such technical ideas as "high-temperature cooling and low-temperature heating". In recent years, technological breakthroughs have been made in air source heat pumps, medium-deep geothermal heat pumps and other key equipment, which get increasingly wide application. The latest development of heat pump technologies applicable to public buildings will be introduced specifically in the subsequent chapter about key technologies. As to the energy use of cooking, fossil fuels are still used in some public buildings, but can be replaced through effective electrification. Moreover, compared to other environments with scattered cooking, including residential buildings, the electrification of cooking in public buildings is easier and faced with less resistance in terms of the adaptation to changes in traditional cultural habits in residential buildings. For instance, cooking has been fully electrified at the newly-built Beijing Daxing International Airport, and the electrification of cooking has been vigorously promoted in many malls and commercial complexes as well. In terms of domestic hot water, the total operation energy consumption of a centralized domestic hot water system is generally 3–4 times the required heat from hot water consumption at terminals, because most of the heat is lost in circulating pipelines' heat dissipation and circulating pumps, which leads to lower use intensity at terminals and reduced overall efficiency of the system. By contrast, a distributed mode of domestic hot water supply makes it convenient to use various heat pump water heaters. Further adoption of distributed domestic hot water systems in public buildings is expected to enable a significant decline in unwanted heat losses and the improvement of overall system efficiency. Steam is needed in hospitals and other special occasions. However, the traditional mode of centralized steam conveyance has such shortcomings as serious waste during conveyance. At present, technologies for their electrification using heat pumps have been developed.

All of the above measures have provided favorable conditions for full electrification first in public buildings.

In future low-carbon energy systems, the role of buildings will change significantly. Buildings will change from pure energy users to complexes integrating energy generation, consumption, and regulation and storage, and from pure energy consumers using rigid loads to participants deeply involved in the construction and regulation of low-carbon energy systems and in the transition to flexible loads. They will become an important link in future low-carbon energy systems. Utilizing renewables of buildings is also an effective means to reduce carbon emissions in the building sector. The current development of PV technologies has provided an important guarantee for the large-scale utilization of PV power in buildings, and the rooftops and external surfaces of buildings are important available space resources. PV power generation and other renewable energy utilization technologies will be better used in public buildings. Although the energy use issues of some high-rise and super high-rise buildings are difficult to solve through utilization of PV, many public buildings with a small shape coefficient have an inherent advantage in full utilization of PV. For instance, transportation buildings including airport terminals and high-speed railway passenger stations have large-area rooftops and few floors, and can achieve a high cost-performance ratio in the laying of PV, see Fig. 4.3. PV power generation has been well applied in completed new station buildings such as Xiong'an Railway Station. The maximized utilization of PV and precious external area of buildings may be considered in more public buildings, whether reconstructed or newly built.

Public buildings will be effective carriers making full use of renewable energy as they have facilitated the self-consumption of self-generated renewable energy, see Fig. 4.4. Let's take typical office buildings for example. After the laying of PV for power generation, the PV power usually can address only about 20% of the peak demands of buildings. However, the characteristics of energy use time of public buildings including office buildings are highly similar to those of the power generation curves of PV and other renewables, so the buildings' PV potential can be fully utilized and renewables of buildings are easy to be totally consumed. The external renewable power including PV power can be consumed effectively as well, thus

(a) Xiong'an Railway Station

(b) China Pavilion of the Beijing International Horticultural Exhibition

Fig. 4.3 Examples of public buildings using PV

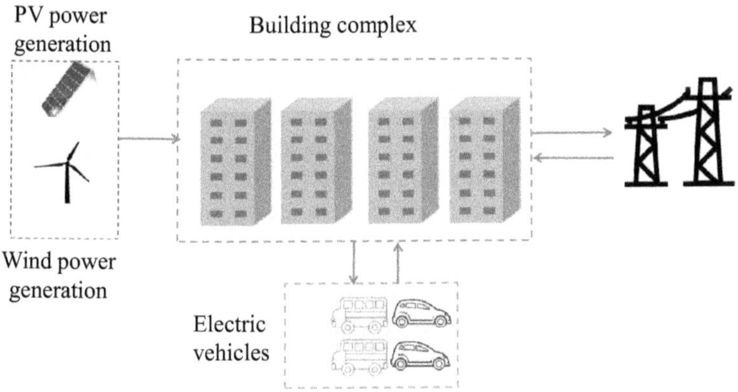

Fig. 4.4 Interaction between public buildings and future low-carbon electric power system

minimizing the increase in energy accumulation and storage costs resulting from the mismatch between the supply of renewable power and the terminal building demand. As public building envelopes have certain thermal inertia and contain numerous energy consumption equipment, a response mode of terminal energy use for adjusting the power of electrical equipment based on the electricity supply characteristics and in proper combination with the energy storage characteristics of building envelopes may be developed in the future, which can further combine active systems, including chilled water cool storage and ice cool storage, to realize flexible energy use of public buildings.

In addition, distributed power storage can be provided in public buildings, and electric vehicle resources in parking lots of public buildings can be properly utilized via intelligent charging piles, which are important approaches to effectively regulate energy use. As the office hours of the personnel is highly consistent with the utilization time of parking lots in public buildings such as office buildings, those buildings have a unique advantage in using electric vehicles in their parking lots to develop new electric vehicle charging piles and charging and discharging modes suitable for PEDF system. The development, popularization, and application of this system mode will contribute to not only the low carbon emission and flexible regulation of energy consumption in public buildings and even the building sector, but also integrate the supply of renewables with energy consumption of building and transportation, achieving low-carbon development of the transportation sector.

The building energy use system is easy to be fully electrified in the first place and managed in a unified manner. This offers favorable conditions for its future involvement in the regulation of the entire energy system and for its response to the changes on the low-carbon energy supply side in the future. Due to the small regulation capacity of a single building, all single buildings need to be aggregated to achieve a large scale before adopting virtual power plant technologies for sound interaction with power grids. In the future, public buildings may be involved in the interaction with power grids in the form of individual buildings and building complexes (building districts), and their electrical loads will shift to flexible loads that respond to the power grid dispatching demands and will allow for flexible regulation of energy consumption within a given scope. Eventually, the interaction between building complexes and power grids will enable buildings to consume surplus renewable power from power grids during troughs of electricity consumption and to transmit electricity to power grids during peaks, thus reducing stresses on power grids. Achieving this goal requires reform of the energy use pattern, the means of energy accumulation and storage, and other aspects of public buildings, as well as the support from new technological means and key equipment.

4.2 Are Public Buildings Still Need to Be Energy-Saving in Addition to Low-Carbon?

Energy demands of public buildings can be reduced through energy saving. Based on lower energy demands and through further electrification, it is possible to achieve low carbon and carbon neutrality goals. On the contrary, if the base of energy demand is huge, it will be difficult to achieve the same. For example, the current annual average electricity consumption of public buildings in China is mostly 50–100 kWh/m^2. The constraint and guide values of energy consumption (non-heating) of several common types of public buildings listed in the *Standard for Energy Consumption of Building* (GB/T 51161-2016) are mostly at the above level, and only the energy consumption of some special public buildings such as large business supermarkets is higher. By contrast, the average electricity consumption of public buildings in the US is mostly above 200 kWh/m^2. China is undergoing high-speed social and economic development. In the future, whether to maintain the current level or allow for a substantial increase in the energy use intensity of public buildings, as the US and other developed countries have done, will determine whether public buildings can achieve the low carbon goal or not. Saving energy and reducing unnecessary energy demands can lay an important foundation for the low-carbon development of public buildings. Further adjustment of energy structure and adoption of effective flexible regulation measures based on energy saving will be conducive to the low-carbon development of the entire energy system.

Take a look at the energy supply side. The zero-carbon electric power system mainly consists of such basic power sources as nuclear, hydroelectric, wind and solar PV power. In the future, electricity generated by fossil energy such as thermal power will become a peak-shaving resource in the zero-carbon electric power system. China's resource status makes it very difficult to have more than 700 million kW of nuclear power and hydropower, leading to a generating capacity of less than 4 trillion kWh (200 million kW of nuclear power and 500 million kW of hydropower; refer to Sect. 2.1), and the insufficient part of power can be supplied only by wind and solar PV power. However, wind and solar PV power cannot be regulated and controlled, so matching flexible power sources are required for coordination. The currently available zero-carbon flexible power sources cannot be anything but hydropower and pumped energy storage, which are the best flexible power sources for coordination with wind and solar PV power. The regulation issue is easy to solve when the flexible power capacity required for the total scale of wind and solar PV power is no more than the total capacity of the hydropower and pumped storage power stations. If the total demand is excessive and more wind and solar PV power will be required to meet it, the regulation can only rely on chemical energy storage, hydrogen storage, air compression and other high-cost energy storage modes, thus causing the cost per unit of energy to skyrocket. Increased energy costs will lead to not only higher living costs but also grown manufacturing costs and weakened competitiveness. This in turn will have a great impact on the national economy. **Hence, to achieve low-carbon development for the whole society, energy saving is still critical and should be prioritized at any time**.

The relationship between "low carbon" and "energy saving" should be viewed dialectically as they are highly uniform in some cases but have certain differences in other cases.

From the perspective of goal, the goal of low carbon in the building sector is to reduce carbon emissions in the life cycle of buildings through effective measures, including carbon emissions during construction, from building materials and during building operation. Carbon emissions from construction activities usually have a low proportion in the total. Developing various low-emission building materials is an important measure to reduce carbon emissions from building materials. As a major part of carbon emissions in the life cycle of buildings, carbon emissions during building operation are mainly from energy consumption (direct carbon emissions from the burning of fossil fuels and indirect carbon emissions from electricity use and other forms). In general, reducing energy consumption can promote the reduction of carbon emissions during operation.

The fundamental goal of building energy saving is to reduce energy consumption during the operation of public buildings by measures including the reduction of energy consumed in cooling, heating and electricity power. The energy saving of public buildings has been carried out for over 30 years in China. With China's rapid urbanization, tremendous changes have been seen in the quantity and quality of public buildings and qualitative leaps in energy saving through positive efforts and creative works of relevant sectors under the guidance of central and local construction authorities. In the 1980s and 1990s, the main problems of China's public buildings were the inadequate quantity and inferior service level, and they were unable to achieve continuous quantity increase and quality improvement as required by social development and the improvement of people's living standards. At that time, China was in the early stage of urbanization and had no adequate experience in constructing large-scale modernized public buildings, so large-scale public buildings were designed and constructed with reference to foreign experiences. Many technologies now widely applied in public buildings were introduced at that time as advanced technologies from developed countries, including large glass curtain wall technology, the primary-secondary pump mode for air-conditioning chilled water, and the pressure-independent variable air volume (VAV) system. Moreover, to ensure construction quality and enable buildings to meet the basic energy saving requirements, China has organized technical experts from all sectors to design standards for public buildings including building energy saving standards since the 1990s. In 1993, China issued its first energy saving standard for civil buildings (GB50176-93), which made comprehensive provisions on the thermal insulation properties of envelopes and the window-wall ratios in different climatic zones. The standard explained to the society "how" to implement energy saving of buildings for the first time and promoted the development of China's building energy saving from the field of scientific research to large-scale engineering application. The *Design Standard for Energy Efficiency of Public Buildings* (GB50189-2005) issued in 2005 played a huge role in standardizing China's public building energy saving. The standard specified detailed design methods of public building envelopes, electromechanical equipment, electromechanical system forms and parameters, and other parts closely related to building energy saving. Over the years, the idea of "speaking with real data" in the energy saving of public buildings has been enjoying popular support, and energy saving has shifted from being measure-oriented to result-oriented, The effect of building energy saving is judged from the final results of energy consumption. The *Standard for Energy Consumption of Buildings* (GB/T 51161-2016) issued in 2016 measured the energy saving effects of buildings with final energy consumption data, and provided an important basis for standardizing the operation energy consumption and furthering reasonable energy saving of civil buildings in China.

Starting with the actual result-oriented idea above, the research on low carbon is still in its infancy compared to building energy saving; however, by leaning the experience from building energy saving, the actual effects and carbon emissions can be regarded as a measure for the final low carbon effect. To achieve the low carbon goal, attention should be paid to the actual carbon emission data in each phase, based on which the carbon emission level should be measured. For the purpose of reducing final carbon emissions, corresponding standards and policies should set such indicators as the reference value or recommended value of carbon emissions. The achievable actual effects should be regarded as important characteristic parameters of low carbon/carbon emission technologies that can be adopted in the construction and renovation of public buildings.

From the perspective of phases, both low carbon and energy saving are important goals that need to be noted in the life cycle of buildings. In the design phase, the focus should be put on how to select suitable low carbon/energy saving technologies. More and more building design methods developed currently are seriously selecting reasonable technologies in the design phase to achieve favorable operation effects of buildings. In the design process, it should be noted that a completed building must be able to achieve the low carbon/energy saving effects during its operation instead of being a simple patchwork of technologies. The constraints in achieving low carbon/energy saving goals in the operation phase should be considered to better the design and make it truly oriented toward the low carbon and energy saving goals. In the operation phase, reasonable operation regulation strategies should be followed to exert the effects of active and passive technologies in buildings sufficiently, so that the building energy use system can operate in the right state to achieve energy-saving/low-carbon operation of buildings.

From the perspective of measures that can be adopted, energy saving of public buildings is expected to be achieved through the technical means for effective energy consumption reduction, design of reasonable systems, improvement of energy utilization efficiency, use of efficient processing equipment, and development of reasonable operation regulation measures. Possible carbon emissions from public buildings include carbon emissions in the life cycle from construction, building materials and operation process. In the construction process, carbon emission reduction should be considered and the main approaches include carbon reduction technologies. On the basis of carbon emission reduction, selected building materials should also meet building's functional requirements and the energy saving needs in the later operation process. In the operation process, possible carbon emissions include direct carbon emissions resulting from the burning of fossil fuels and indirect carbon emissions from the use of external electricity. Avoiding the burning of fossil fuels is the basic measure to achieve the low carbon goal of public buildings. In addition, further changing the energy structure and reducing carbon emissions from energy use are also significant actions to achieve the low carbon goal. For example, in terms of how to achieve the low-carbon heating goal of buildings, the analysis of existing studies indicates that it is very difficult to find lots of available zero-carbon heat sources to fully meet building heating demands, so it is necessary to significantly reduce building heating demands by effective building energy saving measures. In this way, the demands for zero-carbon heat sources can be reduced, thus laying a favorable foundation for establishing the entire low-carbon heating system.

According to the above analysis, the energy saving of public buildings plays an important part in promoting low-carbon operation of public buildings. Energy saving is the foundation, and avoiding unnecessary waste of energy and reducing unnecessary energy use demands are the basic conditions for establishing a reasonable building energy use system. On this basis, the adoption of effective energy saving approaches and measures will contribute to the reduction of building energy demands and carbon emissions. From the perspective of renewable energy available to public buildings, if public buildings use their own PV, the PV usually can meet less than 20% of their electricity demands only, and it is very difficult to address their total energy demands by laying a large area of PV in them. In this case, reducing energy demands or consumption of public buildings through effective energy saving approaches is still the most basic way. Reducing energy demands of public buildings and realizing energy saving provide the basic conditions for reducing the demands for external energy and also help to realize the low carbon goal of buildings.

The difference between the low carbon and energy saving goals is that the annual total energy consumption of buildings was regarded as the criterion when the focus was only on building energy saving before. When low carbon is set as a goal, the relationships among energy structure form, energy demand and energy system supply shall be further considered. From the perspective of building energy use structure, low carbon has set new requirements, that is, the use of fossil fuels in buildings shall be avoided whenever possible and full electrification shall be realized. Carbon emissions in the phase of public building operation include direct carbon emissions resulting from the burning of fossil fuels and indirect carbon emissions from the use of external electricity. Avoiding the burning of fossil fuels is the basic measure to achieve the low carbon goal of public buildings. In addition, further changing the energy structure and reducing carbon emissions from energy use are also significant actions to achieve the low carbon goal. Energy structure adjustment is important to the achievement of carbon neutrality. The building sector shall actively adapt to the significant changes in the future energy structure by adjusting the energy use structure of buildings and avoiding direct carbon emissions from the burning of fossil fuels whenever possible. This lays an important foundation for energy saving and carbon emission reduction of buildings.

In terms of the relationship between building energy demand and external power grid supply, the matching of building energy demand to external power supply on a time scale is required, in order to achieve the low carbon goal. Solar PV power, wind power and other renewable energies, which provide an important foundation for the achievement of carbon neutrality, have high volatility, and their changes can hardly match the building energy demand completely. When the energy demands of buildings are met through sufficient use of their own PV or external renewable powers, there will be a mismatch or inconsistency between renewable power and building energy demand, causing the low-carbon operation and energy saving goals of buildings to be no longer strictly consistent and even to be inconsistent in some cases. For instance, from the perspective of energy saving, such technologies as ice/chilled water cool storage may lead to increased system operation energy consumption instead of reducing it. However, from the perspective of the low carbon goal, cool storage technologies are important measures that can be considered to realize flexible building energy use as they can help to level out peaks of energy use loads in buildings. In addition, in terms of energy saving, the direct conversion of high-grade heat into low-grade heat via direct electric heating is the "use of high-grade energy for low-grade applications". However, in terms of low carbon, this system form can also be adopted if the electricity for electric heating comes from building PV, and the electric heating, by ingenious design including building energy storage (such as radiant electric heating under the concrete floor), can consume PV power in real time and also meet the requirements for time stability in building heating. This is based on the understanding that buildings, as an important part of the entire energy system, can help the energy system achieve the low carbon goal if they become flexible loads that can be regulated, instead of the traditional understanding that buildings are only energy consumers and only need to save energy in themselves and reduce their energy consumption.

It is very difficult to totally match the changing curve of wind and solar PV power with the building electricity consumption curve due to the inconsistency in time. This mismatch between supply and demand may be alleviated through regulation by certain energy use equipment (Fig. 4.5). Although this energy use regulation strategy seems insufficient for energy saving from the perspective of electricity consumption, it is intended to enable the entire energy system to achieve the low

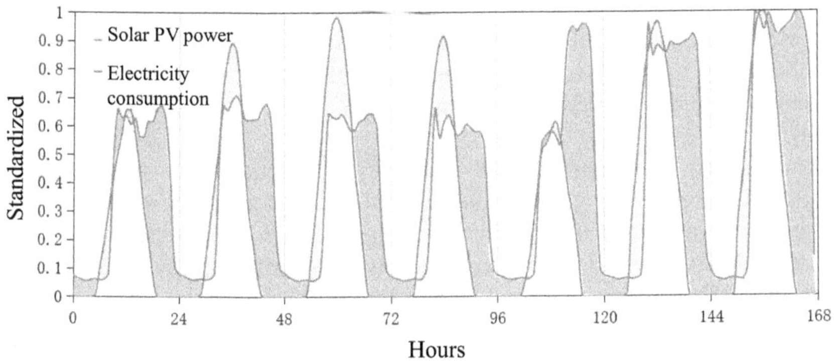

Fig. 4.5 Mismatch between PV renewable energy and building energy consumption

carbon/zero carbon goal in a better way, make buildings effective renewable energy consumers and promote the coordination in the entire energy system from production to consumption.

When the renewable power from buildings themselves or from external sources is higher than the building energy demands during actual operation, the buildings need to consume such renewable energy in a timely manner and utilize the renewable power that does not match in time. If it is necessary to further consume more renewable power, such renewable energy can be consumed effectively by proper increase of energy use and through the use of active energy-using equipment. For instance, if PV generation output is high and the corresponding energy demand of buildings is low in a day with strong solar radiation, the renewable power generated by buildings may be higher than their electricity consumption. If there is still available renewable power when cool/heat/electricity storage and other technologies for buildings have been totally put into use or have reached the upper limit of storage, this part of electricity can be consumed by active consuming equipment in the building. The variable frequency operation of common equipment such as air conditioning units, water pumps and fans may enable certain power regulation, thus adapting to such excessive supply of renewable energy within a short time. In this case, although the energy consumption of buildings is not low enough, renewable energy has been fully utilized in the entire system and the low carbon goal of the building energy use system has been achieved.

If the renewable power from an external power grid needs to be consumed in addition to PV and other renewable powers from buildings, the equipment system in a building can be fully utilized for power regulation in a short time, so as to promote the consumption of renewable energy from an external grid. For example, when the wind power of an external power grid peaks in winter, the air source heat pump in a building can be used to accumulate a certain amount of energy. Such renewable power can be consumed by turning on the heat pump units, and the energy use periods can be shifted based on the energy storage characteristics of building heating, thus allowing for a better match between the building energy use and the supply of renewable power including wind power to certain extent. Although this response and dispatch method may make the total electricity consumption higher than that in the method of supply on demand and on time, it is still beneficial to the achievement of low carbon and full utilization of renewable energy in the entire energy use process.

Consequently, the carbon neutrality goal drives the organic unification of the low carbon and energy saving goals for public buildings. Energy saving is an important foundation for achieving the low carbon goal, while the low carbon or even zero carbon goal sets new and higher requirements for building energy saving. In the future low-carbon energy system dominated by renewable energy, energy supply will face the important task on how to respond to the characteristics of changes in renewable energy, and the building energy use shall shift from setting simple energy demands and acting as energy consumers to adapting to the changes in the supply side, adapting to the future energy system that supplies renewables by proceeding from the demand side, and establishing new supply and demand relationships. The new power distribution systems of PEDF buildings to be introduced later in this report are intended to realize flexible energy use of buildings, thus better helping the entire energy system or even the whole society achieve the low carbon goal.

4.3 CO_2 Concentration and Ventilation for Epidemic Prevention

(1) **Inspection of CO_2 concentration during indoor environment monitoring and assessment**

CO_2 concentration, an important indicator to reflect indoor air quality or freshness and the impact of outdoor fresh air on indoor air, shall not be too high or too low in order to create a suitable indoor environment. According to existing results of CO_2 concentration tested in specific places, CO_2 concentrations in different buildings currently exhibit the following three major conditions:

- **First, CO_2 concentration is lower than 600 ppm**, which is close to that in an outdoor environment (the CO_2 concentration in an outdoor environment is mostly about 400 ppm). For instance, in spite of the influence of such factors as passenger flow changes, the measured CO_2 concentrations in public areas of a lot of metro stations indicate that they are mostly in the range of 400–600 ppm in station halls and platforms and much lower than the limit value (1,500 ppm) set in the design code for metro. CO_2 concentrations in station halls are slightly higher than those in the platforms. The distribution of CO_2 concentrations in many airport terminals also shows that their CO_2 concentrations are only about 600 ppm.
- **Second, indoor CO_2 concentration is in the 600–1,200 ppm range**, with a certain difference from outdoor CO_2 concentration. As revealed by test results of numerous office buildings, even if the mechanical fresh air system is not turned

on, the CO_2 concentration in an office is only about 700– 800 ppm in general and seldom exceeds 1,000 ppm due to the impact of air infiltration into buildings. For instance, the daily CO_2 concentrations in a typical room of an office building mostly change in the 500–800 ppm range in different seasons.

- **Third, indoor CO_2 concentration is over 1,200 ppm and even exceeds 2,000 ppm**, with a difference of over 700–1,000 ppm or even higher from outdoor CO_2 concentration. As to some meeting rooms and classrooms, relevant studies show that their CO_2 concentrations increase obviously over time. The CO_2 concentration in a meeting room or classroom has little difference from the outdoor concentration and is only below 500 ppm upon the entry of personnel. It rises obviously as relevant activity (meeting or lecture) is carried out, and can exceed 1,000 ppm or even 2,000 ppm upon the end of the meeting or lecture.

The reasons can be further analyzed based on these findings. For buildings in normal operation or use, a lower CO_2 concentration means that the difference between indoor and outdoor environments is smaller, thus in the first condition mentioned above, the "effective fresh air volume" is often excessive and the supplied fresh air volume is much greater than the actual demands of personnel. The reason for this condition may be that there are too few people indoors (fewer CO_2 sources). A more important reason may be that the fresh air entering rooms through multiple channels including infiltrated fresh air and mechanical fresh air is excessive and much more than the actual demands of indoor personnel, resulting in a lower indoor CO_2 concentration. Such a large volume of ventilation may cause huge cooling and heating loads in summer and winter, except for spring and autumn when good natural ventilation is expected in buildings. For example, in some airport terminals[2], as the heating load caused by air infiltration accounts for over 50% or more of the total heat supply of an air conditioning system in winter, relevant measures should be sought to reduce the excessive fresh air volume in such occasion and to increase indoor CO_2 concentration from less than 600 ppm to 800–1,200 ppm. For the third condition mentioned above, the indoor environment may be affected by poor ventilation or insufficient ventilation volume to meet the fresh air demands of the crowds indoors in a short time. Relevant measures shall be taken to increase the ventilation volume.

Figure 4.6 shows the relationship between indoor CO_2 concentration and per capita fresh air volume (with very light personnel activities and an outdoor concentration of 400 ppm). According to the figure, indoor CO_2 concentration is about 1,000 ppm when the per capita fresh air volume is about 30 $m^3/(h{\cdot}person)$. The lower the CO_2 concentration is, the bigger the fresh air volume for every 100 ppm change in the CO_2 concentration will be. For buildings in normal operation or use, a lower indoor CO_2 concentration (for instance, 600–700 ppm) means that the difference in CO_2 concentration between indoor and outdoor air is small (only 100–200 ppm). In this case, the fresh air volume in buildings is much greater than the actual demands of people, and corresponding measures shall be taken to reduce the excessive fresh air entering rooms (such as minimizing air infiltration by reducing mechanical air volume or by closing vents). When indoor CO_2 concentration is more than 1,200 ppm, the difference in CO_2 concentration between indoor and outdoor air will be over

700 ppm. In this case, the fresh air volume in buildings cannot effectively meet the demands of people and hence should be properly increased. In this way, similar to indoor temperature and humidity control, indoor CO_2 concentration shall also be controlled within a reasonable range. An excessively low indoor CO_2 concentration means that the effective fresh air volume may be excessive, which may lead to huge air conditioning loads in summer and winter. An excessively high value means that the indoor fresh air volume is not enough to effectively eliminate CO_2 generated by indoor personnel. In general conditions, the fresh air volume is in direct proportion to the energy consumption for the handling and conveyance of fresh air, while the relationship between the capability of fresh air to dilute indoor pollutants and the fresh air volume is nonlinear. As shown in Fig. 4.6, the indoor pollution is reduced to half the original level when the per capita fresh air volume increases from 10 m^3/(h·person) to 30 m^3/(h·person). However, the indoor pollution is reduced only a little when the per capita fresh air volume grows from 40 m^3/(h·person) to 60 m^3/(h·person). Therefore, in the case of a slightly high indoor CO_2 concentration, the indoor air quality can be improved effectively with increased ventilation, without consuming too much energy. On the contrary, in the case of a very low indoor CO_2 concentration, the indoor air quality cannot be improved much with further increased fresh air volume, but too much energy will be consumed. As a result, the CO_2 concentration inside buildings shall be controlled within a suitable range (such as 800–1,200 ppm) to effectively avoid the deficiency caused by an excessively high or low concentration, make sure the indoor environment is in a proper range, and avoid excessive energy consumption resulting from fresh air. So, indoor CO_2 concentration shall be treated as an important basis in the design, operation and regulation of indoor environment systems of buildings.

What we pursue is "effective fresh air" rather than "identifiable fresh air". Both the outside air infiltrated into buildings through gaps in doors and windows and opened exits and entrances and the outside air introduced into buildings by active means including mechanical ventilation fans or air handling units are fresh air, which can effectively influence the indoor environment, meet the fresh air demands of personnel,

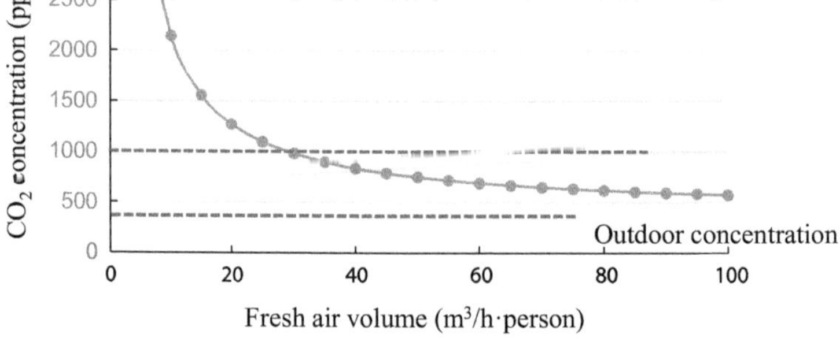

Fig. 4.6 Relationship between indoor CO_2 concentration and per capita fresh air volume

and play an effective role. At present, health and epidemic prevention authorities insist on checking whether the identifiable fresh air volume is up to standard, that is to say, checking the mechanical fresh air volume that can be measured directly is regarded as the basis of whether the fresh air volume in buildings meets demand. This is a reason for the excessive fresh air volume in many buildings. If only mechanical fresh air or such "identifiable fresh air" is accepted, it is impossible to consider the impact of actual air infiltration into buildings.

Now that the purpose of guaranteeing proper fresh air volume is to effectively eliminate the pollutants released by personnel and other indoor pollution sources, why not directly monitor indoor CO_2 concentration to check against the elimination of pollutants generated by indoor personnel or the freshness of air? This is similar to heat and humidity removal, which is intended to guarantee that the indoor temperature and humidity are at a suitable level. Although heating and humidity loads need to be calculated in the design of a heating, ventilation and air conditioning (HVAC) system, indoor temperature and humidity, instead of cooling and heating capacity, must be checked to make sure they meet standards during system operation. Temperature and humidity are the most direct and convenient measurable and controllable indicators for indoor environment, while such parameters as load and cooling and heating capacity are not the most direct or convenient indicators acting on the indoor environment. Therefore, air quality or fresh air shall be treated in the same manner. The purpose of guaranteeing proper fresh air volume is to ensure the health demands of indoor personnel, the required freshness of indoor air and the required indoor air quality. During operation, it is very difficult to reflect indoor fresh air volume or quality comprehensively by measuring "identifiable fresh air" (mechanical fresh air volume) only. Instead, monitoring indoor CO_2 concentration can directly serve as an important means to indicate whether indoor air quality or freshness is up to standard. In the past, real-time field measurement of CO_2 was of high cost and the reliability of instruments was poor, thus limiting the operability of treating CO_2 as a target parameter for direct control and management. Now, the measurement by CO_2 sensors has no significant difference from temperature and humidity measurement in terms of cost and reliability. Therefore, we should adapt to new changes, and regard field CO_2 measurements together with temperature and humidity as main parameters for operation management and for the evaluation of indoor environment control results. Paying attention to and stressing CO_2 concentration rather than per capita fresh air volume will enable more genuine representation of indoor environment conditions, more scientific guidance on HVAC system design, and more convenient system operation and regulation. Taking indoor CO_2 concentration instead of per capita fresh air volume as an important indicator to depict indoor environment conditions is an important embodiment of creating suitable indoor environments truly based on actual demands.

Figure 4.7 lists the common measurable and adjustable parameters and indicators in the creation or control of indoor environments of buildings, which mainly include the above-mentioned temperature, humidity and CO_2 concentration. The goal in the creation of indoor environments of buildings is to control these important indicators

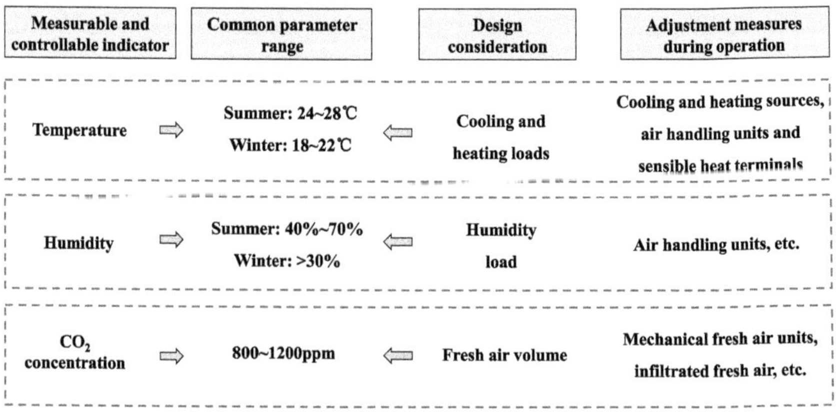

Fig. 4.7 Main indicators considered in indoor environment monitoring and operation

within a suitable range. These indicators should not be too high or too low. Excessively high temperature and humidity in summer will make people uncomfortable, while excessively high CO_2 concentration means that indoor air is of poor quality or not fresh. Excessively low temperature and humidity in summer will lead to a significant increase in cool loss in air conditioning systems, while excessively low CO_2 concentration means that the fresh air volume is more than the actual demands of people and will also lead to unnecessary cool loss or heat consumption in air conditioning systems. To achieve the above-mentioned control objectives, cooling and heating and humidity loads as well as the fresh air volume available to the inside of buildings (demands are to be determined based on the number of personnel, and air volume includes mechanical fresh air volume and air infiltration) should be considered during system design. In actual system operation, indoor environment parameters shall be adjusted and controlled in accordance with measurable temperature, humidity and CO_2 concentration; sensible heat terminal devices and cooling and heating source equipment shall be adjusted to meet temperature control requirements; air handling units for dehumidification or humidification shall be used to adjust indoor humidity; monitored indoor CO_2 concentration shall be used to adjust the fresh air volume when necessary; possible adjustment measures shall be taken for mechanical fresh air and infiltrated air to maintain indoor CO_2 concentration within a proper range (e.g. 800–1,200 ppm).

Hence, in the design and operation of indoor environment control systems for buildings, CO_2 concentration shall be targeted, treated as a control indicator as important as temperature and humidity control, and maintained in a reasonable range, neither too high nor too low. Required fresh air volume shall be calculated based on the required CO_2 concentration and the possible CO_2 amount from indoor personnel and other sources to finish system design. During operation, CO_2 concentration instead of mechanical fresh air volume shall be inspected directly. If monitored indoor CO_2 concentration is below ~800 ppm, indoor CO_2 concentration shall be adjusted through the reduction in mechanical fresh air, and in the fresh air introduced into buildings via building envelopes and other means. If monitored indoor CO_2 concentration is above ~1,200 ppm, indoor CO_2 concentration shall be adjusted through the increase of mechanical fresh air volume or the proper increase of the volume of fresh air introduced into buildings through vents on envelopes. Thus, for the indoor environment control or creation process, the indoor environment creation requirements can be met more effectively when temperature, humidity and CO_2 concentration are taken as important measurable and controllable parameters or indicators. Fresh air volume or mechanical fresh air volume will no longer be emphasized purely or unilaterally. Instead, such parameters or indicators will be used to reflect the genuine conditions of indoor environments and whether the actual indoor fresh air volume can meet the demands of personnel and other sources. Indoor environments will be created in a more scientific and reasonable manner accordingly.

(2) **Independent control of multiple parameters by multiple means**

Technical solutions vary with indoor environment creation requirements, so different paths shall be adopted rather than coupling different requirements together for a joint solution. Figure 4.8 shows the main indoor heat, humidity and pollution sources affecting building environments and the relations between them and the outdoor ones. The thought and solution for independent control of parameters can be developed to control indoor environments. Considering the different air purification, temperature and humidity requirements, the main task in the indoor environment creation process will be how to effectively keep such indicators within a suitable range including indoor temperature T_r, humidity d_r, inhalable particles, and CO_2 concentration c_r as a sign of indoor chemical pollution. Solutions may vary based on different sources and different task requirements:

- **Temperature control**: It is in essence the process of heat transport. Building heat sources include personnel, equipment, lights, solar radiant heat entering rooms and heat transferred by envelopes. Heat supply in winter is actually intended to

Fig. 4.8 Relationship of heat, humidity and pollution sources between indoors and outdoors

make up for the shortage of heat resulting from heat dissipation of envelopes or from air infiltration. In accordance with the characteristics of these heat sources, air conditioning in summer can be accomplished in different ways. Heat can be transported or removed by radiation, natural convection or forced convection. With the idea of air conditioning with independent temperature and humidity control, adopting high-temperature cold sources for indoor heat transport is an important approach to the improvement of energy efficiency during temperature control or heat transport. Furthermore, indoor heat sources differ in temperature and grade. Theoretically, different grades of cold sources can be utilized for heat transport and temperature control. Up to now, some scholars have proposed the improvement of heat removal efficiency through the use of grade features of such heats as the heat transferred by envelopes and through such heat removal methods as the cooling water from cooling tower.

- **Humidity control**: This process needs to be realized by the diffusion effect among air of different moisture contents. For air conditioning in summer, it is necessary to introduce dry air into rooms and remove moisture from indoor humidity sources; in winter, humidification is required due to excessively low moisture content in indoor air. With certain air handling devices, air can be handled to the air supply state required in winter and summer respectively, meeting the humidity control requirements.
- **Control of indoor chemical pollutants including CO_2 and VOC**: All of these pollution sources are generated indoors. The feasible way is to dilute them with outdoor air through ventilation. That is to say, the control of indoor chemical pollutants mainly depends on ventilation.
- **Control of indoor inhalable particles including dust and PM2.5**: Such pollutants as particles and dust may come from the outside (for instance, PM2.5 out of limit) or be generated indoors. Filtration is the most effective way to remove such pollutants. Currently, there are two ways of filtration. One is recirculating filtration of indoor air, namely, filtration by the so-called room air purifier. The other is filtration of fresh air through the fresh air system. When an independent fresh air fan or fresh air handling device is installed, although it is capable of effectively purifying and filtrating fresh air sent into a room and then sending the purified air into the room, outdoor air does not enter the room through the mechanical fresh air system alone as there are many other passageways in a building. In fact, there is abundant unorganized air infiltration in buildings. Such air infiltration can also carry dust from the outside to the inside, which cannot be filtered and handled by filters installed in the fresh air system. Therefore, filtration alone cannot ensure effective removal of PM2.5 from fresh air. Moreover, fresh air system reduces PM2.5 and other contents in fresh air through filtration, but the removed pollutants are accumulated in the fresh air handling device. As it is impossible to clean fresh air filters every day, secondary pollution will be caused when outside clean air passes through them. Thus, "indoor low pollution is caused when there is high pollution outdoors, and indoor low pollution is still there when there is no pollution outdoors. " Upon installation of indoor independent purifiers, the air volume

to be filtered can be selected according to indoor purification demands, thus realizing rapid indoor purification. When both the outside and the inside are clean without pollution, no filtration will be required, and thus no secondary pollution will be caused. Moreover, even if outside air with high concentrations of particles or dust and PM2.5 enters the inside, larger particles and dust in it will adhere to indoor surfaces due to gravity and can be effectively removed by such measures as sweeping instead of becoming a load for the indoor circulating purifier. By contrast, if the removal of pollutants is totally dependent on the filtration of fresh air, the aforementioned pollutants including large particles and dust will become major loads for fresh air filters. In addition, the indoor circulating purifier can also capture dust and particles generated indoors to ensure that indoor PM2.5 concentration is at an appropriate level. The indoor filter can only filter indoor particles (which may be from the outside or generated indoors), without any influence on the control of temperature, humidity and chemical pollutants.

- **Virus etc.**: Main transmission routes of the COVID-19 include contact, droplet and aerosol (airborne) transmissions. Ventilation requirements are increasingly stricter under the impact of COVID-19. The virus can be considered a type of particle. To prevent the potential risk of airborne transmission, measures similar to tackling particle pollution and ensuring cleanliness shall be taken. Clean air (virus-free air), which is fundamental to achieving adequate dilution and reducing the risk of airborne transmission, can be obtained through the introduction of outside air/fresh air (which can be considered virus-free) and adequate purification of indoor air. If indoor air is used as clean air, the efficiency of purification must be sufficient to effectively intercept and remove potential virus particles in indoor air.

Thus, through the analysis for different indoor environment parameters or different control targets, it can be seen that indoor temperature, humidity, air quality (CO_2 and other indoor sources), particles and other environment parameters, which are from different sources and of different properties or which may be treated in different ways, should not be controlled through the coupling of temperature and humidity control together or the simple coupling of the control of pollutants from different sources together, but should be controlled by different methods: **(1) Temperature should be controlled through cooling or heating; (2) Humidity should be controlled through corresponding humidification or dehumidification; (3) VOC, CO_2 and other pollutants generated indoors can be replaced only through the introduction of outside low-concentration fresh air; (4) Pollutants including particles can be controlled through the installation of indoor circulating purifiers**. In this way, adopting different treatment methods for different indoor environment control targets is conducive to better meeting the respective indoor environment control requirements.

Meeting the requirements of temperature, humidity and cleanliness control in an indoor environment separately can help to achieve better effects of indoor environment control and avoid deficiencies resulting from mutual coupling and unified control. Let's take the cleanroom of the electronic industry for example. Their process and production environments have very strict requirements on indoor temperature, humidity and cleanliness, and the parameter control requirements are much higher than those in common public buildings. However, there is a significant difference in air volume required for temperature, humidity and cleanliness control. Generally, the circulating air volume passing through high-efficiency filters to meet high-level purification requirements (dozens of and even hundreds of times of air change) is much more than the air volume required for temperature and humidity control (usually one or two or several air changes only). Hazardous gases generated in technological processes can be removed only through the change with outside air, thus there are certain fresh air exhaust requirements. Besides, there are dust sources indoors, which need to be removed via self-circulating purification and filtration. As cleanrooms are required to be ultra-clean, the requirements for particles are very high, leading to the need for a large air volume. Hence, this need is met with self-circulation instead of all fresh air. At present, the form of an environment control system of makeup air unit (MAU) + dry cooling coil (DCC) + fan filter unit (FFU) is usually adopted in such circumstances. Specifically, the MAU is used to handle fresh air sent into cleanrooms to meet the indoor ambient humidity control requirements, and different treatment measures are provided according to the annual changes in fresh air to ensure there is enough exhaust air to remove hazardous gases generated in technological processes. The DCC is used to handle indoor sensible heat loads, and medium-temperature chilled water (~14 °C) is used to meet the indoor temperature control requirements. The FFU is used to filter indoor circulating air to remove pollutants generated in technological processes, thus meeting the indoor cleanliness requirements. As a result, the environment control system for an electronic cleanroom should be a system solution starting from separate control of temperature, humidity and cleanliness. Certainly, when air/air supply is used in practice to meet temperature, humidity and cleanliness control requirements, a small air volume (for example, in many cases, the air volume required for dehumidification is small) can fulfill multiple control tasks at the same time (for instance, the air volume for dehumidification also fulfills certain cooling and cleaning tasks). On this basis, it is necessary to further make up for the insufficient air volume according to control targets to meet all control requirements. The indoor environment parameter control in public buildings is not as strict as that in cleanrooms of industrial buildings. Nevertheless, similar ideas of indoor environment control can be still applied, namely, indoor temperature, humidity and cleanliness shall be controlled separately, respective functions shall be performed in different treatment processes, and countermeasures shall be taken for different control targets to meet indoor environment creation requirements in different aspects respectively.

(3) **Use of clean air to ensure adequate epidemic prevention**

At present, the transmission mechanisms of COVID-19 remain to be further identified. However, according to virus transmission routes and cases of transmission in enclosed spaces as released by authorities and in related studies, the risk of airborne transmission cannot be excluded. According to the *Diagnosis and Treatment Protocol for Novel Coronavirus Pneumonia (Trial Version 8)* of China, "Transmission of the virus happens mainly through respiratory droplets and close contact. There is potential aerosol transmission in a relatively closed environment for extended exposure to high concentrations of aerosol. As the virus can be isolated from feces and urine, attention should be paid to feces- or urine-contaminated environment that may lead to aerosol or contact transmission."

In the face of the COVID-19 pandemic, the possible transmission routes include contact, droplet and aerosol (airborne) transmissions. Aerosol transmission means that the virus can be transmitted in the air in the form of aerosol. Virus particles at a certain concentration exhaled by the infected are likely to be transmitted for a certain distance via air. According to some current study results, certain virus concentrations have also been detected in mouth and nose exhalation of the infected in inpatient wards, indicating that viruses exhaled by the infected probably are spread by air. As airborne/aerosol transmission is possible, it is necessary to respond to such transmission risks in practical epidemic prevention by taking effective blocking measures.

Various epidemic prevention guidelines have mentioned such measures as keeping adequate ventilation and increasing ventilation, which provide an important guarantee for reducing risks of disease spread and avoiding airborne transmission. The starting point for increasing ventilation to reduce airborne transmission risks is that adequate ventilation and introduction of a sufficient amount of outdoor fresh air are considered to be able to cause viruses released by the infected to be diluted adequately and their concentrations to be lowered to a range that is small enough. The outdoor fresh air, namely, virus-free air, is regarded as an effective means of virus dilution. From this perspective, as long as the air is free of viruses and even contains viruses at concentrations much lower than potential concentrations of viruses exhaled by the infected, it can be used as an effective carrier to dilute the potential airflows from the infected. The effective mixing of airflows of different concentrations and different orders of magnitude (the airflow with low virus concentrations or the clean airflow is much greater than the exhaled airflow from the infected) will enable adequate dilution of potential viruses exhaled by the infected to a concentration low enough and a dilution factor low enough as well. In this way, the risk of potential airborne transmission can be effectively controlled and reduced to a level low enough.

The Wells-Riley equation can be used to evaluate the risk of airborne transmission, as follows:

$$P = \frac{C}{S} = 1 - e^{-Iqpt/Q} \tag{4.1}$$

where, P—risk of infection; I—number of infected persons; q—quantum generation rate produced by an infected person (quanta/h); p—respiratory capacity of a person, to be selected according to activity status; t—exposure time (h); Q—room ventilation rate (m^3/h).

According to the equation above, the airborne transmission risk is subject to the combined influence of factors like ventilation rate, exposure time and possible quantum generation rate produced by an infected person (Fig. 4.9). Being essential

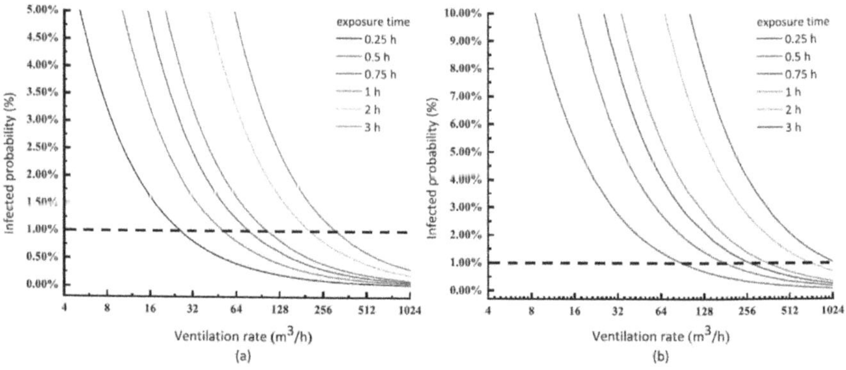

Fig. 4.9 Relation between airborne transmission risk and ventilation rate[3]: **a** $q = 14$; **b** $q = 48$

to the dilution of viruses in a room, ventilation rate Q is also an important means to control pollutants like the novel coronavirus which can be considered particles transmitted by air within an effective concentration range. It needs to be emphasized that, for the dilution of viruses, all pathogen-free air can be used as an effective carrier to remove or reduce pathogen particles in indoor air. In terms of airborne transmission of viruses, such air is also "clean air" that is available and performs the dilution function, in addition to outdoor air. When there is a purification device that can kill pathogens in a room or an air conditioning system, the ventilation rate of the purification device shall also be included. In view of the above-mentioned influencing factors, Q can be calculated with the following equation:

$$Q = Q_f + Q_p \eta_p \tag{4.2}$$

where, Q_f represents fresh air volume (m^3/h); Q_p represents air volume treated in the filter section of a purifier or an air handling unit (m^3/h); and η_p represents first pass efficiency in the filter section of a purifier or an air conditioning unit.

As the atmospheric environment has excellent dilution effect on outdoor air, outdoor fresh air can be considered completely clean and free of virus particles except for in an extremely particular environment. The transmission of viruses in the air can be considered the diffusion of particles and can be treated in ways similar to tackling particulate pollution (by the purification of particles). The roles of purifiers and return air filters depend on the actual setting and equipment parameters. Air filtration, electrostatic precipitation and other indoor particulate removal technologies can effectively reduce indoor particulate concentrations and guarantee indoor air quality. For instance, the filter section is installed in air handling units or air ducts of air conditioning systems to filter and purify air, and individual air purifiers can be used in rooms for effective purification of room air. Therefore, as long as methods that can effectively remove polluting particles with a particle size similar to that of a novel coronavirus are adopted, the goal of lowering novel coronavirus concentrations in air can be achieved. In this case, the purified return air can be used as a carrier for effective dilution of potential virus concentrations. However, it should be specially noted that the selected particulate removal products shall have high efficiency by number, instead of efficiency by weight, in removing particles with a particle size similar to that of a novel coronavirus.

As shown in the above analysis, air free of virus or with virus concentrations much lower than those in exhaled air flows from the affected, whether fresh air or purified indoor return air, can be used as an effective carrier to dilute virus concentration. Thus, in the case of using clean air to reduce the risk of airborne transmission of viruses and meet epidemic prevention requirements, the following two situations should be considered:

- **Areas with potential infection sources are known**. For example, in hospital wards, international arrival areas of airport terminals and sampling sites for nucleic acid testing, the potential infection sources include patients, international arriving passengers and people sampled for nucleic acid testing respectively, and the corresponding staff are the major groups to be protected, which means that the potential infection sources and people in need of protection are identified in these situations. In these areas, ventilation requires the creation of a rational air distribution direction to enable fresh air to flow through the zone of protected groups to the zone of potential infection sources and to inhibit the air from flowing back whenever possible. In this case, all fresh air shall be used (to ensure the air flowing through protected groups is not polluted) and the air from the potential infection sources shall be discharged. The fresh air volume shall meet the air distribution requirements and shall not flow back.
- **Areas with potential infection sources are unknown**. For example, in stations, airports, malls and meeting places, it is very difficult to foreknow potential infection sources and thus impossible to create a definite airflow field by controlling airflow. In this case, the key is to dilute the air as much as possible. Large volume ventilation is required to ensure there is sufficient clean air to protect all people whenever possible. However, the air for ventilation is not necessarily fresh air. Even if conveyed via air ducts, circulating air can also be used as clean air as long as it has been filtered. Larger circulating air volume can guarantee higher clean air volume and lead to a greater dilution factor and better dilution effects. In such areas, it is unnecessary to overemphasize the need to achieve a large number of fresh air changes. Instead, enough clean air volume shall be guaranteed to achieve a dilution factor large enough for effective dilution of potential viruses exhaled by the infected, thus reducing the risk of potential airborne transmission.

CO_2 concentration, as an indicator, can effectively reflect the fresh air volume required by indoor personnel. For the aforementioned target of keeping indoor CO_2 concentration within a certain range (such as 800–1,200 ppm), the corresponding per capita fresh air volume is usually at 20–30 m^3/h. With this per capita fresh air volume, the corresponding dilution factor or infection risk can be calculated when combined with the concentration of viruses exhaled by the infected. The dilution factor that can be achieved with this fresh air volume is usually sufficient to meet the fresh air requirements for reducing virus concentration if effective indoor air distribution is maintained. Further excessive increase in fresh air volume for reducing indoor CO_2 concentration will not lower the potential airborne transmission risk significantly. Therefore, in practice, it is unnecessary to blindly require increasing ventilation. If the ventilation volume is excessive and has obvious adverse impacts on indoor thermal environment in winter, epidemic prevention requirements shall be considered, as appropriate, on the basis of meeting the basic requirements of indoor thermal environment. As long as CO_2 concentration, when taken as an indicator, is not excessively high (for example, not more than 1,000 ppm), the ventilation volume can be considered sufficient to meet the clean air requirements for reducing airborne transmission risk indoors. Besides, CO_2 concentration should not be too low (such as below 600 ppm); otherwise, a large volume of unnecessary outdoor air will be introduced, making it more difficult to regulate and control the indoor thermal environment.

As a result, in the prevention and control of airborne transmission of COVID-19, the situation where areas with potential infection sources can be identified shall be distinguished effectively and handled differently from the situations in which it is impossible to identify areas with potential infected persons. The key to risk reduction is ensuring sufficient clean air volume, which can be achieved through appropriate and reasonable ventilation and installation of scattered purifiers indoors for air purification. For the potential exhalation of viruses by the infected, effective clean air can be used as the dilution carrier to achieve a sufficient dilution factor, thus meeting the requirements of reducing/diluting potential virus concentrations. Therefore, in the control of indoor air quality, environment and epidemic prevention risks, indoor CO_2 concentration should be taken as an indicator to check the actual control results, and controlling CO_2 concentration within a certain range should be treated as an important target in creating indoor environments.

4.4 Further Understanding on the Contribution of Public Building Energy Systems to the "Dual Carbon" Goals

Building energy demands usually involve cooling, heating (including domestic hot water), electricity and some other aspects. The building energy system is a basic part for the building to fulfill the daily operation functions, create a suitable and comfortable environment and ensure normal operation of all energy-using equipment in the building. The fundamental task of the system is to meet the demands for cooling, heating, electricity and other energy sources in the daily operation of the building. For a traditional building energy system, its main objective is to meet the building energy demands. When saving building energy and reducing building operation energy consumption are emphasized, efficient operation and energy consumption reduction for the energy system can be effectively promoted through the adoption of various technological means, such as energy system adjustment and the adoption of

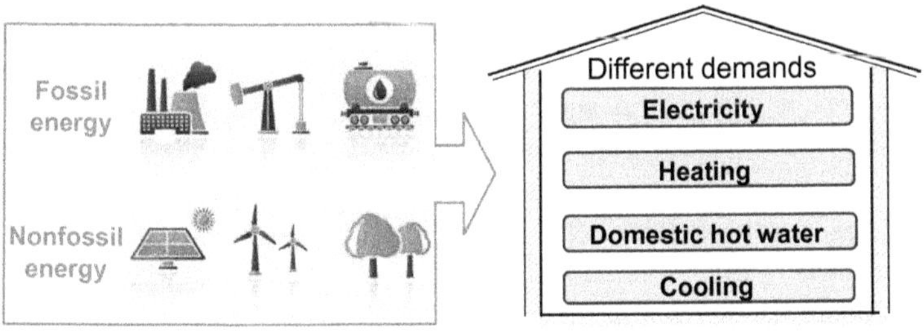

Fig. 4.10 Public building energy demands and main supply modes

high-efficiency energy system equipment. Under the guidance of the "dual carbon" goals, the public building energy system needs to fulfill its fundamental functions. On this basis, it is necessary to reflect on its role orientation in the entire energy system in the future and further understand the corresponding roles and tasks it needs to undertake to achieve the low carbon and even zero carbon goals for the entire energy system when the building changes from a traditional unit of terminal energy use to a complex that integrates energy generation, storage and consumption. This is also a major challenge for the development of a public building energy system.

Among the current energy supply modes corresponding to cooling, heating and electricity demands in China's public buildings, the common modes include electricity and fossil fuels (coal, natural gas, oil, etc.). These energy sources are supplied to buildings and then further utilized or treated by appropriate energy-using equipment in buildings to meet the demands of buildings (Fig. 4.10).

- Cooling is essentially a process of "transporting" indoor heat to the outside (transporting from indoor heat sources to outdoor heat sinks). Implementing this process requires the consumption of work, thus work (exergy) needs to be provided from the outside. Cooling demands are met mainly by electrically driven refrigeration units (including all types of water chilling units and air conditioning units), or alternatively by thermally driven/gas driven absorption refrigeration units (adopted in very few public buildings). Hence, from the perspective of final energy solutions, cooling demands are mainly met in the mode of electric drive.
- Heating is essentially a process of heat transfer (from heat sources to rooms). Heat can be supplied to the inside of buildings after it is obtained from suitable heat sources that have been sought and a certain temperature difference is set for the transfer. Heat is mostly supplied from municipal heating, from boilers set in public buildings, or from heat pumps, namely, heat is transferred from the outside or obtained through the burning of fossil fuels in buildings, or the heat grade is improved and heat is transported by the input work of heat pumps. Driven by the carbon neutrality goal, clean and low-carbon heating has received more and more attention in municipal central heating and other heating modes. Various methods for effective residual heat recovery are adopted to obtain clean heat sources as many as possible for the supply to buildings, thus reducing carbon emissions. In scattered heat supply modes including boilers that burn fossil fuels

in buildings, fossil fuels inevitably lead to carbon emissions, so their availability shall be reconsidered. Various heat pumps often meet building heat demands by input power, and the energy supply mode to meet final demands is still electricity.

- Domestic hot water is a special heat demand for buildings, which is often addressed by the burning of fossil fuels, by solar water heaters, or by electrically driven heat pumps and even electrical heating. The steam needed in some places like hospitals is still obtained mainly by the burning of fossil fuels or electrically driven steam making at present. In some public buildings, the energy demands of cooking are still partly met by fossil fuels (including natural gas).
- Electricity is used to meet the demands of lighting and all types of electric equipment (including heat pumps, water pumps, fans, and other HVAC system equipment) in buildings. The current electric power supply is mostly from the municipal power grid. In the future, with the further development of PV and other renewable energy sources in buildings, external power grid and the building itself will jointly serve as the electric power supply sources for the building. The building energy use system shall adapt to this change in the electric power supply source.

Hence, in terms of the ways to meet the cooling, heating and electricity demands of public buildings and the final modes of energy supply at present, electricity is the leading form of energy, and the direct burning of fossil fuels for heating or cooling is still adopted only in some public buildings. At present, electricity consumption often accounts for up to 50–100% of the terminal energy consumption of public buildings. Under the guidance of the carbon neutrality goal, full terminal electrification is an important goal and guideline in the development of the public building energy system. It is an effective way to avoid the direct burning of fossil fuels at public building terminals that results in direct carbon emissions. Besides, it is also an important channel to realize low-carbon development for public buildings and develop a low-carbon and even zero-carbon energy system.

Considering the common energy supply modes for the public building energy system as mentioned above, fossil fuels are mostly used to meet heating demands and the demands in special occasions. Therefore, using electricity to meet the demands in special occasions and replace fossil fuels is an important approach toward full electrification, which needs the support from appropriate electrically driven equipment. For instance, for such energy-using equipment in public buildings as the boiler for heat production, replacing it with electrically driven modes including the heat pump is an effective approach. All types of heat pump equipment meeting heat production demands are growing vigorously (the development of air source heat pumps will be elaborated in Chap. 5). In hospitals and other public buildings with steam demands, it has been possible to remove fossil fuel boilers and make steam with heat pumps. In some public buildings where gas is needed for cooking purposes, electrified alternatives and corresponding cooking utensils have grown mature gradually. Further development of all types of electrical equipment adaptable to full electrification demands will provide an important support for the full electrification of the public building energy use system and for avoiding the direct burning of fossil fuels in public buildings that results in carbon emissions.

On the basis of full electrification, the challenges facing public buildings are further observed. After full electrification of buildings, the relationship between the change in demand of building energy use (electricity consumption) system and the energy (electricity) supply needs to be further considered, which necessitates a clear understanding of the demand and supply characteristics. The energy use of public buildings usually varies by day, week, month and season, which is closely related to the use modes and operation demands of public buildings and also to the outdoor climate. The energy consumption curve of a public building depends on its internal use functions and electrical equipment as well as the external climate. Take Shanghai for example. The typical daily normalized electricity consumption curves for public buildings in this city are shown in Fig. 4.11. It can be seen that the uptimes of main equipment in all types of public buildings are close (7:00–9:00), but there is a big difference in the downtimes of main equipment due to different operation and use modes in these public buildings. The energy use peaks of buildings generally last 8–9 h. However, the peaks of tourist hotel buildings last longer, up to 12 h, due to their operation characteristics. Moreover, the daily electricity consumption curves of public buildings are close to the daily changing curves of solar PV power generation in shape, thus providing a favorable condition for better consumption of PV power.

Take Shanghai for example. Table 4.2 shows the ratios of total electricity consumption in peak period to that in valley period (peak-to-valley ratios) for public buildings in a day. Overall, the peak values of electricity consumption of public buildings in that day occur at 11:00–12:00, while the valley values at 3:00–4:00. The peak-to-valley ratios for office and mall buildings are larger due to their intermittent operation mode, while those for tourist hotel buildings and medical buildings are smaller as they are often wholly or partly in 24-h continuous operation. The electricity consumption troughs for all the buildings occur in the small hours, thus the daytime peak electrical loads can be shifted to nighttime through electricity storage, ice cool storage and other energy storage technologies for peak-shaving.

On the other hand, the carbon neutrality goal drives the reform of the supply and demand relationship in the future energy system. In the future electric power system

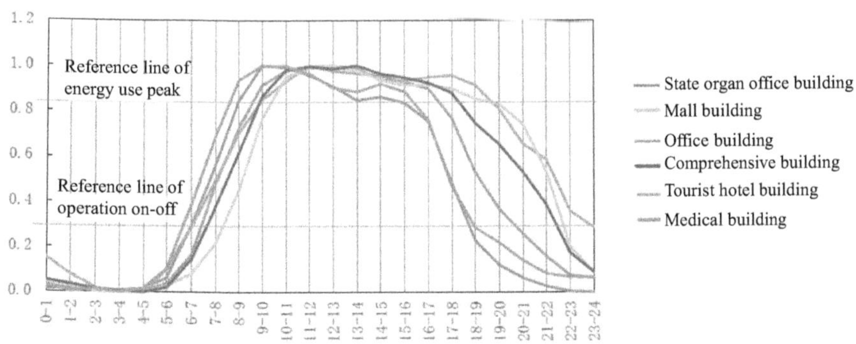

Fig. 4.11 Hourly normalized electricity consumption curves of public buildings in Shanghai (a typical day in summer)[1]

Table 4.2 Peak-to-valley ratios of daily electricity consumption of public buildings in Shanghai (a typical day in summer)[2]

No	Type of building	Peak-to-valley ratio
1	Mall building	3.9
2	Office building	3.0
3	Comprehensive building	2.8
4	State organ office building	2.6
5	Tourist hotel building	1.4
6	Medical building	1.4
7	All connected buildings	3.0

Notes 1. Peak-to-valley ratio = Total electricity consumption of peak periods/Total electricity consumption of valley periods
2. Peak periods: 8:00–11:00, 13:00–15:00, 18:00–21:00; Valley periods: 22:00–6:00 next day

with renewables playing a leading role, it will be difficult to supply wind power, solar power and other renewables which are characterized by natural volatility according to electricity demands of users; instead, it will be necessary to seek an effective approach to adapting to the change and volatility characteristics of the energy supply side and coordinate the supply and demand sides to help to achieve the low-carbon goal. With this trend in development, the electric power system is in urgent need to shift from "source changing with load" to "load changing with source". If energy use loads can match the energy power supply dominated by renewable power, they will contribute to better coordination between supply and demand. Buildings, as major energy use loads, also need to adapt to such development requirements of the electric power system. Thus, from the perspective of electricity supply characteristics, in the future energy supply structure dominated by renewable power, public buildings need to cooperate with the power generation side of power grids in the consumption of renewable power, shift from the traditional mode of "demand-dependent supply" to a new mode of "supply-guided demand", and adapt to challenges posed by the structural reform on the energy supply side. With this development trend, it is of paramount importance to realize flexible electricity consumption of buildings.

Progress is being made in the study on building flexibility and on how to make buildings more flexible both at home and abroad. Grid-friendly buildings, electricity demand side response, building flexibility and flexible building have become hot topics in relevant studies. For example, the International Energy Agency's Energy in Buildings and Communities (IEA EBC) Programme has carried out multiple projects known as Annexes to study flexible buildings, including Annex 67-Energy Flexible Buildings, Annex 80-Resilient Cooling of Buildings and Annex 82-Energy Flexible Buildings Towards Resilient Low Carbon Energy Systems. Up to now, many demand side response projects have been carried out for buildings, including Direct Load Control and Time-of-use Pricing, which have exhibited considerable peak shaving potential in building flexibility/resilience regulation. With regard to building flexibility, there are still many issues to be studied, including how to achieve unified depiction, the establishment of a unified analytical framework, and modeling methods.

If the energy for public buildings is supplied jointly by external electric power and renewable power (such as PV) generated in such buildings themselves on the basis that such buildings are fully electrified, buildings will become an important link in the entire energy system, thus it will be of paramount importance to make them more flexible and achieve the flexible use of electricity. For the new "PEDF" building energy use system to be introduced in the subsequent chapter "Key Technologies", its key or purpose is to achieve "flexibility". Flexible electricity consumption necessitates the adaptation of public buildings to such development requirements and the corresponding reform of energy use characteristics and demands of them. To realize flexible electricity consumption, public buildings shall seek effective energy accumulation and storage as well as the release and use approaches from their building entities, electric equipment and available surrounding conditions (such as electric vehicles), use energy in a more flexible (adjustable) way, and adapt to the development requirements of energy system under the goal of carbon neutrality.

- Building envelopes and internal furniture usually have certain thermal inertia, causing the building to have certain heat storage characteristics in creating a thermal environment. Effective utilization of the energy storage characteristics for different public buildings to enable the buildings to adapt to response requirements of the energy power system under certain operation regulation strategies is an important measure for increasing building flexibility via this passive link. How to depict the thermal inertia of building envelopes and internal furniture, the heat storage capacity, and how to achieve effective linkage between heat storage capacity and building energy use system are issues to be further studied in depth.
- Cool/heat storage methods, as common methods for energy accumulation and storage in buildings, can be used for daily, monthly and even seasonal energy storage. Ice/water cool storage, as common cool storage methods in public buildings, are intended to lower the energy use peak of building air conditioning and to achieve effective load shifting. Cool storage methods will also become an effective means for flexible regulation of energy consumption. Heat storage methods are often used in heating systems or domestic hot water application scenarios. Their heat accumulation and storage capacity is also a very good means for public building flexibility regulation. Common cool/heat storage means are mostly used for daily energy storage and load transfer, while some methods for seasonal energy storage are also under development and are expected to narrow the gap in building energy demands among seasons and to become an effective means for seasonal flexible regulation.
- Electrical equipment is the leading energy consumer in buildings and also an important resource of resilience/energy use flexibility. They are expected to offer great potential of flexible dispatching. For the flexible regulation capabilities and methods for various electrical equipment, it is necessary to further understand the load characteristics of such equipment, which can be divided into turn-off, migrating and power-adjustable loads. For example, the use of some electrical

equipment can be advanced ahead of schedule or postponed within a certain time range, and indoor lighting can have its electrical load adjusted within a certain range of brightness adjustment. More and more variable frequency fans and water pumps are used in public building air conditioning systems. Frequency regulation enables such fans and water pumps to achieve considerable power regulation within a period of time without any obvious impact on the terminal environment, thus it is also a means of flexible regulation. How to make electrical load characteristics of various electrical equipment further adapt to the demands for flexible regulation of building electricity use is also a major requirement proposed for the future development of such equipment.

- The accumulator is an effective measure for building energy accumulation/storage. Although the current accumulators and other chemical energy storage means are still costly and accumulators still present risks, distributed accumulators combined with buildings will remain the most important energy storage means in buildings. The capacity of accumulators to be set in buildings can be determined according to the building energy demands, the building energy use curve expected by electric power system and other available energy accumulation/storage capacity of buildings. The form of accumulator, energy storage capacity, the capability of participating in building flexibility regulation and the response order required in buildings also need to be further studied.
- The electric vehicle, as the backbone of future transportation, is also an important storage battery resource. In the future, electric vehicles will replace today's fossil fuel vehicles, and their battery size and corresponding power storage and discharge capacity will play a significant role in consuming renewable power and promoting the coordination between energy supply and demand. Users of public buildings and those of electric vehicles are almost the same, so the interaction between electric vehicles around buildings and building energy use has significant potential for making building energy use more flexible. This necessitates the full consideration of electric vehicle resources in a public building energy system to enable them to coordinate with the system in achieving effective dispatching and to participate in the regulation of building energy. In this way, the battery resources and the charge and discharge capacity of electric vehicles will be brought into full play.

The development goal for the flexible electricity consumption of public buildings is that the future operation can be based on the electricity consumption curves expected on the external electric power supply side to assist the external electric power supply side in the full utilization of renewable energy and in the achievement of low carbon and even zero carbon, while the building energy use demands are met. The aforementioned "passive" (buildings) + "active" (electrical equipment and energy storage measures in buildings) energy accumulation, storage and utilization and load transfer can facilitate the flexible regulation of energy use demands of public buildings to a certain extent. The combined use of the above flexible measures will enable more flexible building energy use and is also expected to enable buildings, on the basis that their basic energy demands are met, to better respond to the dispatching demands on the power grid side, thus achieving "load changing with source" in a real sense.

Therefore, in terms of the trend for the public building energy system, full electrification or re-electrification is fundamental to ensuring the transformation of energy use structure of public buildings and promoting low-carbon development of energy consumption. For a fully electrified public building energy system or an electromechanical system, corresponding high-efficiency electrification devices and equipment shall be further developed to meet the demands of cooling, heating (including domestic hot water), electricity and other energy sources. In the meantime, in the context of full electrification, buildings will serve as an important link in the entire energy system and need to proactively adapt to the reform of energy system. Targeted at the adjustable flexibility of building energy system, it is necessary to further explore possible scenarios, technologies and applications to achieve friendly interaction between buildings and the entire energy system and to contribute to the low/zero carbon goal in the whole society.

4.5 Further Understanding of District Energy Systems Oriented Toward the "Dual Carbon" Goals

The district building generally refers to campuses, communities, parks and other building complexes composed of individual buildings of various types and functions. A building complex may consist of buildings of a single type, such as multiple office buildings, or consist of multiple types of functional buildings. In terms of the scale of a district, it may cover an area of 1–100 km^2, and the total building area may be 100,000–10,000,000 m^2 or even more. District building energy may be considered as a "source-grid-load-storage-use" system that meets the electricity, air conditioning, heating and other energy demands of the above-mentioned buildings and their related industries.

During district energy planning, designers are very apt to focus on a "centralized" energy system, namely, they would set up an energy station in a unified manner for district cooling and district heating, and even set up a combined cooling, heat and power (CCHP) system. Is it suitable to set up such a system? How to understand such district buildings with the scale and volume much greater than those of common individual buildings, and how to establish a reasonable district energy system? To answer those questions, overall consideration should be taken from such aspects as the asymmetry of "centralization" versus "distribution" and of "cold" versus "heat".

(1) Distribution versus centralization

District cooling or distributed cooling? This is essentially an issue of "centralization" versus "distribution", which has been described at length in the Annual Report on China Building Energy Efficiency of 2014 and of 2018. The respective advantages of a centralized system mode and a distributed system mode are compared from different aspects and angles. The main reasons given by advocates of the centralized mode over the distributed mode are as follows: A centralized mode has a larger scale and higher efficiency; due to benefits of the "coincidence factor", a centralized mode can reduce total installed capacity and investment; a centralized mode saves device space and centralized management can contribute to improved operation management. The reasons for advocating a distributed mode include: Convenient to freely adjust consumption, a distributed mode is capable of flexible response to low utilization rates of 5% or 10%, thus avoiding excessive supply; a distributed mode can flexibly meet the needs of different grade parameters, and the distributed supply avoids the waste of energy quality; a distributed mode saves transmission pipelines and reduces the energy consumption from transmission and distribution.

Whether a distributed mode or a centralized mode is chosen in the creation of a building environment has a significant impact on the energy consumption of the system. As demonstrated by a large number of practical engineering cases, in the face of numerous terminals for different demands, the energy consumption through the adoption of a single centralized system to serve these terminals simultaneously is much higher than that through the adoption of many distributed systems which are independent of each other. A typical case is that the actual energy consumption of air conditioning in a residential building that adopts a centralized air conditioning system is much higher than that in a residential building that adopts a distributed air conditioning system. For office buildings, the energy consumption of an all-air VAV mode, a "fan coil + fresh air" mode and a split air conditioning mode is approximately in the ratio of 3:2:1 given all other conditions are the same; however, for office personnel, there is little difference in their air conditioning effects. If nobody stays in a room, a VAV air conditioning system will still operate, but both a fan-coil air conditioning system and a split air conditioning system can be turned off separately. In case of overtime work in individual rooms in the evening, a VAV system and a fan coil system need to be fully operating, but a split air conditioner can be started separately at will. The reason why a centralized cooling system has higher energy consumption than a distributed system is that multiple demands independent of each other are put together but are different at every moment. To meet the different demands at terminals, a centralized system pays a corresponding "price" for regulation, while a distributed system can better adapt to the changes in regulation.

If there are multiple terminals in a system, the changes in demands consist of two levels: one is the synchronization in "quality" and the other is the synchronization in "quantity". The former is the analysis of changes in terminal demands from the perspective of the grade for a cold/heat source needed by a system (temperature T or temperature difference ΔT), while the latter is the understanding from the perspective of the required quantity of cold/heat Q. When the demands of terminals are out of sync and changes in demands are inconsistent, a centralized system must meet the inconsistent demands of these terminals simultaneously. As to different demands for "quality", for instance, several terminals demand low-temperature (such as 7 °C) cold water and the rest only demand cold water above 10 °C, and a centralized system can only "meet the lower demand instead of the higher demand", which means the supply of 7 °C cold water only, so it is difficult to improve chiller efficiency with increased water temperature. As to different demands for "quantity", when 90% of terminals operate below 5% load and only 1–2 terminals need to be supplied with 100% load, it is very difficult for a centralized system to regulate them all. As a result, insufficient regulation will lead to "excessive supply", or effective regulation will lead to huge energy consumption by the fans and water pumps.

Now that a centralized system has problems as mentioned above, why there are still so many advocates for a centralized system? In general, the reasons are as follows:

1. A larger scale will contribute to a higher degree of concentration and higher efficiency just like the industrial production process. Is this true? For the industrial production process, yes. And it is also true for the energy production and conversion processes, such as the production of coal, oil, gas and electricity. However, different from production, a building is intended to serve its users, i.e. people scattered to different areas of the building. The demands of these users vary significantly in parameter, quantity, space and time, so centralized/unified supply can hardly meet the demands of different individuals. As a result, unified supply often follows the highest demand standard only This is the reason why offices with a central air conditioning system in the US and Hong Kong are rather cold in summer, while many rooms with a centralized heating system in northern China are rather hot in winter. This also leads to the phenomenon that air conditioners of the entire building need to be turned on during overtime of several persons in the evening, and all lights in an open office need to be turned on when only one person feels the need to turn on a light. Energy consumption resulting from such excessive supply is actually much more than that reduced by the high efficiency of the centralized system. Furthermore, large-scale production must be carried out at full load to achieve high efficiency. However, due to the distributed changes in terminal demands, a centralized service system in a building seldom operates at full load but often at partial load and even at a very low percentage of load. Such a low load rate is often caused by some terminals being turned off, instead of the load decrease at all terminals. Thus, the efficiency of a centralized system is low when it operates at a low load rate. On the contrary, a distributed system only turns off terminals not in use, and the terminal load rates of terminals being in use are not low at all, thus the system efficiency is not lowered as well. Unlike the high-efficiency process of mass production with the same parameter in the industrial production process, it is exactly the inconsistency in terminal demand parameters and in time that causes a more centralized system to suffer from lower efficiency.
2. "A more centralized system is easier to maintain and manage?" In fact, operation management covers two tasks: one is the maintenance, management and

repair of equipment, and the other is the regulation of system operation. The former is to ensure safe and reliable operation of each device in the system by timely repair and replacement when any fault is found. The latter is to ensure the system offers the best services efficiently by timely regulation of system operation status according to changes on the demand side. For a centralized system with a small quantity of high-capacity equipment, specialized technicians can be assigned to ensure normal operation of such equipment. A distributed system with numerous pieces of equipment may have a high fault rate, making it difficult to ensure normal operation of the equipment. This probably is another important reason for advocating a centralized system. However, in reality, with the advances in technology, the reliability and level of automation for a single set of equipment have been improved substantially. Besides, the troubleshooting for a faulty piece of equipment can be accomplished through simple replacement in a short time without influencing the normal operation of other equipment. On the contrary, large-scale equipment of a centralized system has a higher fault rate, its fault may have a greater influence, and in most cases, a piece of large-scale equipment that malfunctions can hardly be overall replaced and its on-site maintenance takes an extended period of time. In terms of operation regulation requirements, for a centralized system, it is necessary to ensure the normal operation of each set of equipment. In addition, the transmission and distribution system needs to be adjusted, so that it can change the distribution of circulating water volume, circulating air volume and fresh air volume according to changes in terminal demands. The cold/heat source equipment also needs to be adjusted, so that it can constantly adapt to changes in terminal demands. All of these are important tasks for adjusting the operation of a centralized system. The bigger the system is, the more complex the regulation process will be. Conversely, the operation regulation of a distributed system is very simple. It only needs "on" and "off" actions depending on terminal demands or corresponding regulation of relevant quantity, without the need to adjust the distribution in transmission system. The current automatic control technology can fully meet the control and regulation demands of various distributed systems. The operation practices by the overwhelming majority of distributed systems have also demonstrated their advantages in operation regulation. In this case, does the argument that "a centralized system is easy to maintain and manage during operation" no longer hold up? With the development of information technology, "distributed" systems that adopt digital communication technologies for direct management and regulation of devices distributed everywhere have gradually become the mainstream. "IoT", "Sensor Network" and other emerging technologies in this century have made the management and regulation of distributed systems increasingly feasible, reliable and inexpensive. From the perspective of maintenance and management as well as operation regulation, the trend in the development of building service systems in the future will be increasingly focused on distribution rather than centralization.

3. "Many new technologies are only suitable for centralized systems and new technological development necessitates the development of centralized systems."

Indeed, such methods as ice cool storage and water cool storage are only applicable to large-scale centralized systems. Water source heat pumps and ground source heat pumps also need to be installed in systems of a certain scale. The CCHP system adopting distributed energy technologies is more in need of a centralized system large enough for coordination. If the energy saving effects achieved by the excellent performance of these new high-efficiency energy saving technologies can offset the increased energy consumption resulting from a centralized system, adopting a centralized system to achieve the final energy saving goal is certainly justifiable. However, if the energy consumption increased by adopting a large-scale centralized system is higher than the amount of energy saved by these new technologies and eventually leads to an increase in the actual total energy consumption, then why should a centralized system be selected for the sole purpose of adopting new technologies? As demonstrated in surveys and analyses of practical cases, when a building cooling, heating and power (BCHP) system is adopted, the gas-to-electricity conversion efficiency in the electricity generation part is 40%, which is 15% lower than that of a large-scale gas and steam combined cycle pure power plant (55%). The cooling capacity generated simultaneously by the residual heat of the BCHP system is only 45% of the input gas energy at most, while the same cooling capacity can be generated with less than 9% of electricity according to the current centrifugal chiller efficiency. However, electricity generation is reduced by 15% in the BCHP system in order to generate such cooling capacity. Therefore, the BCHP system is inefficient in terms of energy conversion and full utilization. Moreover, in terms of energy structure, the burning of fossil energy like gas is not conducive to low-carbon development, and it is not advisable to develop a gas-fired CCHP system in the context of "dual carbon" goals. As natural gas also falls into fossil energy, the limited resource of natural gas in the future will be mainly used for peak-shaving of electric power systems. After a CCHP system is adopted, the operation time required for electric peak shaving is different from the time required for cooling and heating, thus leading to the decision on whether to adopt an operation mode of "determining electricity by heat" or "determining heat by electricity". Neither of the operation modes would be suitable if a gas-fired power plant should fully cater for the peak-shaving demands. This is also the reason why CCHP shall not be advocated any more in the context of low-carbon energy development. In this case, selecting a large or even giant centralized system for the sole purpose of applying these "new technologies" obviously does not make much sense. Of course, some public buildings such as jumbo airports, station buildings and large public venues are unlikely to adopt a distributed system due to their nature. Their architectural forms and functions determine that they must adopt a centralized service system. Accordingly, it is justifiable to select and use some new technologies like ice cool storage and water cool storage to support such centralized systems. In practice, not all new energy-saving and high-efficiency technologies are oriented toward centralized systems. To adapt to distributed service modes and characteristics, recent years have witnessed the successive generations of many new technologies and products targeted at distributed systems. A typical successful case is the VRF air conditioning system. Namely, a split air conditioning system is expanded to a multi-split air conditioning system, which not only maintains the characteristics of split air conditioners being distributed and adjustable separately, but also reduces the number of outdoor units and overcomes the difficulty in arranging outdoor units of a split air conditioning system. As a result, a multi-split air conditioning system has become an effective alternative to a conventional central air conditioning system in office buildings. Similarly, large-bay office buildings have adopted a digital lighting control system for separate control of each light. Adopting new technologies to support the idea of an independently adjustable distributed system has proven to be a great success.

The selection between centralized and distributed system modes depends on different demands. In the face of production objects that can be reproduced on a large scale, the industrial production process prefers a large-scale "centralized" mode to achieve high efficiency and low energy consumption, thus a centralized system mode would be adopted in most cases. The building service system serves individual service objects that differ sharply. Multiple relatively independent demands are put together but are different at every moment. Hence, this does not apply to the industrial production mode. The distributed system and the centralized system adopt different ways to address demand characteristics. A distributed system meets the "quality" and "quantity" demands separately by distributed methods, while a centralized system meets the "quality" and "quantity" demands from different terminals by certain regulation and distribution measures. When there are marked differences in the demands from different terminals and inconsistencies in the changes of demands, a centralized system will result in certain losses due to its regulation and distribution methods.

Therefore, **"centralization" and "distribution" are not opposite, but represent a continuously changing process and a game process between "demand" and "supply"**. Whether a centralized system can be adopted and the difference in energy use between a centralized system and a distributed system can be judged through the evaluation on the degree of synchronization in terminal demand changes under all circumstances. When the demands of multiple terminals are obviously different, a centralized mode will cause significant unevenness in regulation, increased losses of cooling/heating capacity, and unnecessary dissipation. In this case, a distributed acquisition mode should be adopted to meet their respective demands. Therefore, in the case of inconsistent changes in the loads of multiple terminals, a distributed system tends to meet the different demands of these terminals more easily than a centralized system, thus avoiding excessive supply. When terminal demands are severely out of sync, the advantages of a centralized system including high energy efficiency will be offset by the energy consumption/loss caused by the huge differences among terminals. This leads to the phenomenon that the actual energy consumption of a centralized system is much higher than that of a distributed system in certain circumstances. For district buildings, it is difficult for the demands of terminal users to be in sync and to change consistently. More often, there are multiple users with distributed and independent demands at one single terminal. Hence, during the establishment of a district energy system, such distributed and independent user demands shall be taken into full consideration. In addition, the terminal user regulation capability shall be brought into full play to enable users to adjust the terminal. In this way, it will be possible to better adapt to the terminal adjustability and reduce the unnecessary waste resulting from a centralized and unified creation system for heat environment.

(2) District energy planning

From the view of characteristics of the demands from district-level users, the building demands usually include cooling, heating and electricity demands, which should be met by external supplies (including the utilization of renewable energy from buildings themselves).

- For building heating, if there exist reasonable heat sources, heat can be obtained through effective heat transfer or from heat pumps to meet demands. If there are external free residual heat resources or concentrated thermal resources available, they can be effectively utilized on the basis that such utilization can bring certain economic, environmental and resource efficiency, so as to meet the heating demands of buildings. For example, in the thermal resource construction for centralized heating in northern urban areas of China, all kinds of industrial residual heat, power plant residual heat and other resources have been collected and utilized effectively through technological innovation. This lays an important foundation for meeting the urban heating demands. District users can also select such heat as an appropriate technical solution to meet their heating demands. **Why is there centralized heating? Because there are residual heat sources, which would be discharged for nothing if there is no centralized heating. Full recovery and utilization of all kinds of low-grade residual heat is the only reason for the development of centralized heating. Centralized heating should not be developed at all if there are no residual heat resources!** When a building needs to seek a reasonable heating solution by itself as there are no external direct thermal resources available, renewable resources including air sources and geothermal resources may be considered. Heat pumps can be used to offer the temperature difference required for heat transport, so heat is transported to meet heating demands. There are some district buildings that obtain heat by the burning of traditional fossil energy including gas and coal. However, this method has been made inadvisable upon adjustment of the entire energy structure and upon the control of carbon emissions in the context of "dual carbon" goals.
- For building cooling, input work (chiller) is required to meet cooling demands and to transport heat. Building cooling and heating are different in nature. The former is a heat transport process that requires input work, while the latter is a heat transfer process. Practical operation effects demonstrate that a district centralized cooling system established without differentiating cold from heat obviously consumes more energy than a distributed system does. **Why not advocate centralized cooling? Because there are no residual cold resources!** The justification or root cause for district centralized heating is that there are residual heat resources

available, which can be collected and transported over a certain distance for effective recovery and reuse. However, for cooling, it is very difficult to find sufficient residual cold resources available to meet the demands of large-scale district cooling. Common outdoor cold sources/heat sinks include air, soil, surface water, rivers and lakes, which can hardly be utilized on a large scale for district-level centralized cooling. On the other hand, the temperature grades of such natural cold sources/heat sinks often can hardly be utilized directly for heat transport at the air conditioning and cooling level. Thus, mechanical work and input work must be used to transport heat and thus meet cooling demands. A district-level centralized cooling mode is inappropriate through overall consideration of such aspects as initial cost and economical efficiency.

- The electricity demand of buildings will become the most important energy demand in the future. Driven by full electrification or re-electrification, electricity will become the leading and even the only energy supply mode in buildings. In addition to the use of electricity for daily lighting, equipment and sockets in buildings, electricity can also be used as the final mode of energy supply for cooking, cooling and heating demands in future buildings (Fig. 4.12). Electrically driven heat pumps will serve as an important approach to meet the cooling and heating demands of buildings. All of these make it extremely important to know or plan for the electricity demands of district buildings.

The energy demands of a district energy system include cooling, heating and electricity demands. From a comprehensive perspective of the characteristics for energy supply and demand as well as of the available low-carbon energy, it is hard to find available cold sources or cooling capacity inputs from the outside. Perhaps available residual heat and other thermal resources might be out there, but electricity is the main energy supply mode available. Hence, the energy planning for district buildings shall focus on electricity planning (including the energy use for building heating) and heating planning. Traditional district energy planning is only targeted at a single energy system, such as the planning for electricity, air and cooling (heating) respectively, leading to artificial separation of the optimal configuration of resources among different energy systems. Facing the future low-carbon energy system development

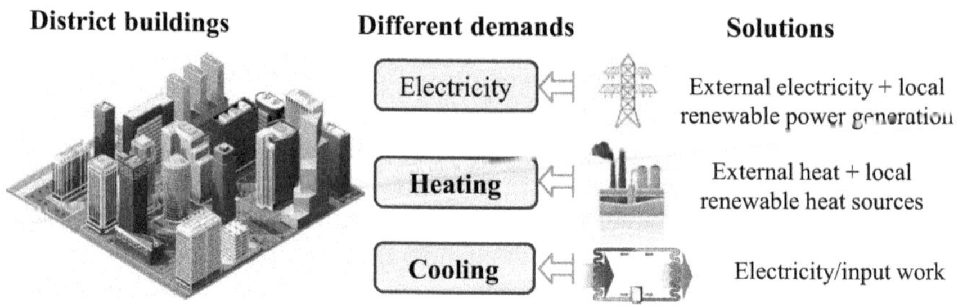

Fig. 4.12 Cooling, heating and electricity demands of district buildings and main forms of external inputs

and energy structure reform, **we need an energy supply support pattern that takes cooling and heating demands into full consideration while focusing on electricity and heating planning based on the concept of collaborative planning of district integrated energy systems**. This pattern shall give full play to the complementary natures and synergistic effects of different energy forms, enable the optimal configuration of resources in the energy system on a larger scale, improve system flexibility and the consumption capacity of renewables, and enhance integrated system energy efficiency.

(3) Overall impacts on the demand side of district buildings

In the face of the "dual carbon" goals, district-level consumers have the potential to become important resources of dispatching response for power demand in future electric power systems. **The marginal effects of improving individual building energy efficiency shall be exhibited gradually. The energy consumption structure on the building user side shall be optimized gradually. Electric energy will become the leading energy type in future buildings**. In the future, the district building will become a key link that plays an important role of regulation in the entire energy system. The scale and electricity consumption of a single building can hardly play a certain role in the power grid dispatching system. However, in the case of expansion to a district, the building scale will become larger and the adjustable capacity of electrical loads will become stronger. As a result, such users of district-level electrical load are likely to be regarded as a kind of effective and dispatchable resources of flexible load for the power grid, which, under the goal of low-carbon development for the future electric power system, will assist the power grid in achieving better dispatching and in adapting to the development direction of the future system dominated by renewable power including wind and solar power.

For district buildings, unified consumption of their own renewable energy may be taken into consideration. With regard to available PV, wind power and other renewable powers in buildings, the energy/electricity demands of single users are out of sync and inconsistent. After such buildings shift to multi-user district buildings, their energy use patterns are expected to better adapt to the characteristics of changes in the supply of renewable energy in the district. The first step is to consume and utilize their own renewable energy. On this basis, the load regulation capability/energy use flexibility that district buildings can achieve is a link that should be further stressed. In the future system of low-carbon energy structure, if building users can respond according to the characteristics of changes on the energy supply side and make their own energy demands better adapt to the characteristics of changes in renewable power supply, they will bring significant benefits to the establishment of the entire low-carbon energy system and to the carbon neutrality goal for the whole society. Due to the energy use characteristics of district buildings, certain energy storage measures may be considered to make their energy use more flexible, for example, adopting certain cool storage (water cool storage or ice cool storage) and heat storage measures. At present, energy storage has become an important measure for many district buildings to regulate loads and reduce peak energy demands. The approach of distributed battery may be further considered as well to increase the flexibility in their energy use, thus better responding to the characteristics of changes in electric power supply. District buildings will enable these energy storage/accumulation measures to have better scale effects in the effective leveling of fluctuations in energy demands and in the better adaptation to external supplies.

In the future low-carbon electric power system, the terminal electricity demand side is expected to respond to the characteristics of spontaneous fluctuation of wind and solar power, etc. on the renewable power supply side; besides, the user side including buildings will be required to become flexible loads. The scale of district buildings will make themselves the main users in the electricity retail market. The district building energy system, as an important application scenario for demand response, will be taken as a crucial means of interaction between supply and demand to facilitate the coordination and interaction between generation-side and demand-side resources (Fig. 4.13). For instance, a district building energy system can adopt effective measures of load regulation to better adapt to the changing demand on the external electric power supply side through the regulation of its own cooling, heating and electricity demands and through certain energy accumulation/storage methods. Future district-level users or district buildings, including large-scale building complexes and all types of campuses, can have their adjustable loads packed to form a converged virtual power plant for overall participation in electricity demand response; moreover, the total electricity consumption can take part in electricity market transaction.

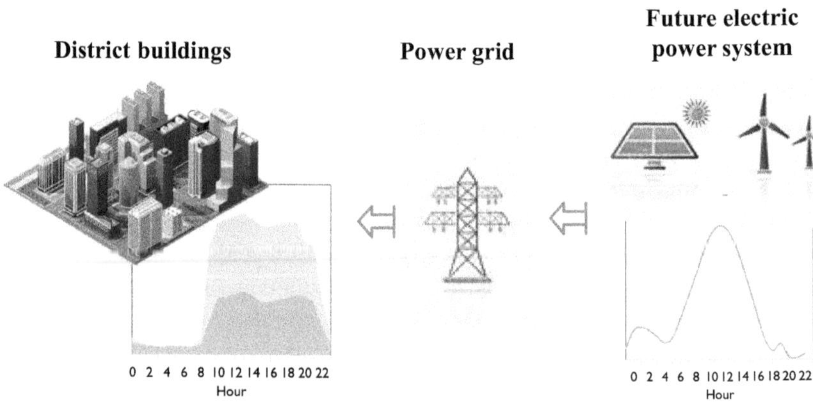

Fig. 4.13 Future district buildings to be involved in power grid response as flexible load

www.ingramcontent.com/pod-product-compliance
Ingram Content Group UK Ltd.
Pitfield, Milton Keynes, MK11 3LW, UK
UKHW021500031026
15588UKWH00002B/912
9789811955273